RESCUED BY

C++
Third Edition

By Kris Jamsa, Ph.D.

gpc

JP

JAMSA
P·R·E·S·S ®
...a computer user's best friend ®

Published by
Jamsa Press
2975 S. Rainbow Blvd., Suite I
Las Vegas, NV 89102
U.S.A.

http://www.jamsa.com

For information about the translation or distribution of any Jamsa Press book, please write to Jamsa Press at the address listed above.

Rescued by C++, Third Edition

Printed in the United States of America.
98765432

ISBN 1-884133-59-2

Publisher	**Technical Advisor**	**Director of Publishing Operations**
Debbie Jamsa	Phil Schmauder	Janet Lawrie
Content Manager	**Cover Photograph**	**Cover Design**
Todd Peterson	O'Gara/Bissell	Marianne Helm
		James Rehrauer
Composition	**Illustrators**	**Copy Editors**
Eugene Marks	Eugene Marks	Ann Edwards
James Rehrauer	James Rehrauer	Dorothy Oppenheimer
Nelson Yee	Nelson Yee	Renée Wesberry
Proofers	**Technical Editor**	**Indexer**
Jeanne K. Smith	Dr. Robert Barger	John Bianchi
	LingYan Tang	

Jamsa Press is a wholly-owned subsidiary of Gulf Publishing Company:

Gulf Publishing Company
Book Division
P.O. Box 2608
Houston, TX 77252-2608
U.S.A.

Table of Contents

V

LEARNING THE BASICS

In this section, you will learn basics you need to know to create your own C++ programs. If you have never created a program before, don't worry, this section starts at step one. To start, you will install the Borland *Turbo C++ Lite* compiler, which this book's companion CD-ROM contains. Using the *Turbo C++ Lite* compiler, you can create, compile, run, and debug (remove the errors from) your own C++ programs. By the time you finish the simple lessons this section presents, you will be well on your way to programming in C++! The lessons in this section include:

SECTION ONE

LESSON 1

INSTALLING THE BORLAND TURBO C++ LITE COMPILER

As you will learn in Lesson 2, "Creating Your First Program," to program in C++, you place your program statements into an ASCII source file. Then, using a special program called a *compiler*, you convert the C++ program statements, which you understand, into the ones and zeros the computer understands and can execute. Today, professional programmers buy and use such C++ compilers as Borland C++ or Microsoft *Visual C++*. To help you get started with C++ programming, Borland International has let Jamsa Press include the *Turbo C++ Lite* compiler on the CD-ROM that accompanies this book. As you will learn, the *Turbo C++ Lite* compiler is an MS-DOS-based program. If you are using Windows, you can run the *Turbo C++ Lite* compiler within a DOS window. In this lesson, you will learn how to install and use the *Turbo C++ Lite* compiler. By the time you finish this lesson, you will understand the following key concepts:

- The *Turbo C++ Lite* compiler is an MS-DOS-based compiler.

- Using the *Turbo C++ Lite* compiler, you can create and edit your C++ program statements, saving the statements to a source file on your disk.

- After you create your C++ source file, you can use the *Turbo C++ Lite* compiler to compile your program statements into the ones and zeros the computer can execute.

- Using the *Turbo C++ Lite* compiler, you can compile most of the programs this book presents.

- Unlike other C++ compilers, such as Borland C++ or Microsoft *Visual C++*, when you compile a program using *Turbo C++ Lite*, you can execute the program that the compiler creates only from within the *Turbo C++ Lite* environment.

- To create an executable program file that you can give to another user to execute, you must compile the program with such compilers as Borland C++ or Microsoft *Visual C++*.

INSTALLING THE TURBO C++ LITE COMPILER

As briefly discussed, *Turbo C++ Lite* is an MS-DOS-based compiler. If you are using Windows, you can run the *Turbo C++ Lite* compiler from within a DOS window, as shown in Figure 1.1.

Figure 1.1 Running the Turbo C++ Lite compiler from within a DOS window.

3

To install the *Turbo C++ Lite* compiler from within Windows, perform these steps:

1. Insert the CD-ROM that accompanies this book within your CD-ROM drive.

2. If you are using Windows 95, select the Start menu Run option. If you are using Windows 3.1, select the Program Manager File menu Run option. Windows, in turn, will display the Run dialog box, as shown in Figure 1.2.

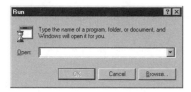

Figure 1.2 *The Run dialog box.*

3. Within the Run dialog box, type in the program name *D:\SETUP*, replacing the drive letter D with the drive letter that corresponds to your CD-ROM drive. For example, if your CD-ROM drive is drive E, type *E:\SETUP*. The *SETUP* program will display the Destination Path dialog box, as shown in Figure 1.3.

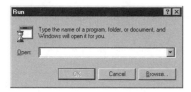

Figure 1.3 *The Destination Path dialog box.*

4. The Destination Path dialog box lets you specify the directory on your hard disk into which *SETUP* will install the *Turbo C++ Lite* compiler. By default, *SETUP* will install the files into the *TCLITE* directory. To use the *TCLITE* directory, select OK to continue. Otherwise, type in the pathname you desire and press ENTER.

You are now ready to run the *Turbo C++ Lite* compiler as discussed in the section titled *Running Turbo C++ Lite*.

INSTALLING TURBO C++ LITE UNDER MS-DOS

As you have learned, *Turbo C++ Lite* is an MS-DOS-based compiler. If you are using MS-DOS and not Windows, perform the following steps to install *Turbo C++ Lite* on your hard disk:

1. Insert the CD-ROM that accompanies this book within your CD-ROM drive.

2. Use the following *XCOPY* commands to copy the *Turbo C++ Lite* files from the CD-ROM onto your hard drive. Within each command, replace the drive letter D with the letter that corresponds to your CD-ROM, as shown here:

```
C:\> XCOPY  D:\BGI\*.*      \TCLITE\BGI\*.*     <ENTER>
C:\> XCOPY  D:\BIN\*.*      \TCLITE\BIN\*.*     <ENTER>
C:\> XCOPY  D:\INCLUDE\*.*  \TCLITE\INCLUDE\*.*  /S  <ENTER>
C:\> XCOPY  D:\LIB\*.*      \TCLITE\LIB\*.*     <ENTER>
```

RUNNING THE TURBO C++ LITE COMPILER

The *Turbo C++ Lite* compiler is an MS-DOS-based compiler. If you are using Windows, you must open a DOS window to run the compiler. To open a DOS window under Windows 95, perform the following steps:

1. Select the Start menu Run option. Windows will display the Run dialog box.

2. Within the Run dialog box, type *COMMAND* and press ENTER. Windows, in turn, will open a DOS window.

To open a DOS window under Windows 3.1, perform the following steps:

1. Select the Program Manager File menu Run option. Windows will display the Run dialog box.

2. Within the Run dialog box, type *COMMAND* and press ENTER. Windows, in turn, will open a DOS window.

Within the DOS window, type the following *CHDIR* command to select the *\TCLITE\BIN* directory:

```
C:\WINDOWS> CHDIR  \TCLITE\BIN  <ENTER>
C:\TCLITE\BIN>
```

Next, type the *TC* command to start the *Turbo C++ Lite* compiler:

```
C:\TCLITE\BIN> TC  <ENTER>
```

Your system, in turn, will start the *Turbo C++ Lite* compiler, as shown in Figure 1.4.

Figure 1.4 The Turbo C++ Lite compiler.

The *Turbo C++ Lite* program normally displays a code window within which you type your program statements, an output window within which *Turbo C++ Lite* will display your program's output, and a message window within which *Turbo C++ Lite* will display messages (such as syntax-error messages that tell you have mistyped a program statement). If you do not see the program, output, and message windows, select the *Turbo C++ Lite* Window menu and choose the Tile option. If after you tile the windows you do not see a specific window, select the Window menu and click your mouse on the option that corresponds to the window you want to display.

LOADING A C++ PROGRAM

In Lesson 2, you will learn how to type your own C++ program statements within the *Turbo C++ Lite* program window. To save time as you examine the programs this book presents, you can load the program files from this book's companion CD-ROM. Using the File menu Open option, you can load a file from the CD-ROM.

The CD-ROM organizes the program files by lessons. For example, the CD-ROM stores the Lesson 1 programs in a folder named *LESSON01*. Likewise, you will find the Lesson 2 files within the folder named *LESSON02*. Finally, you will find the files for Lesson 40 within the *LESSON40* folder. For example, the *LESSON01* folder contains the program file *Demo.CPP*, which contains the following C++ programming statements:

```cpp
#include <iostream.h>

void main(void)
  {
    cout << "This is a sample C++ program!" << endl;
  }
```

To load these program statements into *Turbo C++ Lite*, perform the following steps:

1. Within *Turbo C++ Lite*, select the File menu Open option. *Turbo C++ Lite*, in turn, will display the Load a File dialog box, as shown in Figure 1.5.

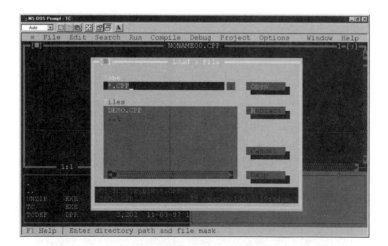

Figure 1.5 The Load a File dialog box.

2. Within the Load a File dialog box, type the filename *D:\LESSON01\Demo.CPP*, replacing the drive letter D with the drive letter that corresponds to your CD-ROM drive. *Turbo C++ Lite*, in turn, will display the program statements within its program window, as shown in Figure 1.6.

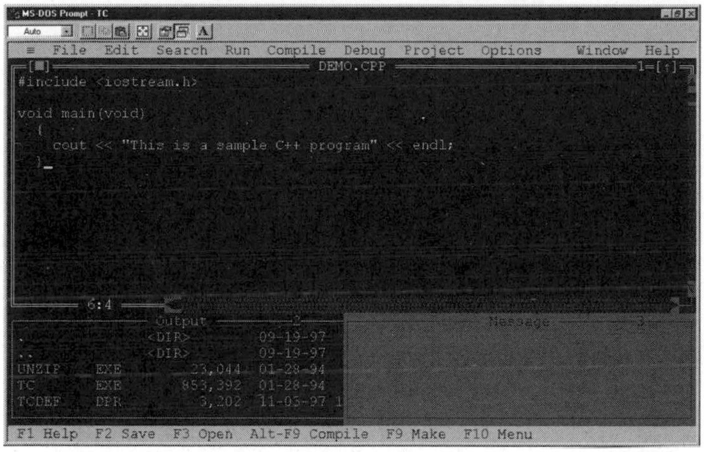

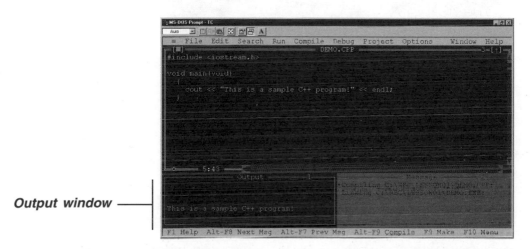

*Figure 1.6 Loading the **Demo.CPP** program into **Turbo** C++ **Lite**.*

Note: *To print your program statements, select the **Turbo** C++ **Lite** File menu Print option.*

RUNNING YOUR PROGRAM

When you program, you type C++ program statements which you later store within a *source file*. Then, using a special program, called the compiler, you convert the program statements into the ones and zeros the computer can execute. To compile your program statements within *Turbo C++ Lite*, select the Compile menu Make EXE File option.

As you will learn in Lesson 2, when you create a C++ program, you must follow the C++ syntax rules, which specify, for example, that you must place a semicolon at the end of each statement. If you violate a syntax rule, your compiler will display syntax-error messages. When a syntax error occurs, you must edit your program statements to correct the error. If, however, your program statements are correct (they do not violate any C++ syntax rules), the C++ compiler will create a program your computer can execute. Within *Turbo C++ Lite*, you can then run the program by selecting the Run menu Run option. *Turbo C++ Lite* will display the program's output, as shown in Figure 1.7.

Output window ⎯

*Figure 1.7 The **Demo.CPP** program output.*

YOU CAN RUN THE PROGRAMS YOU COMPILE WITH TURBO C++ LITE ONLY WITHIN TURBO C++ LITE—NOT FROM A COMMAND LINE

As you have learned, the C++ compiler converts your C++ program statements into ones and zeros the computer can execute. The compiler stores the executable program within an *EXE* file, such as *Demo.EXE*. When you create an executable program using *Turbo C++ Lite*, you can run the program only from within the *Turbo C++ Lite* environment (using the Run menu Run option). If you try to run the program from the system prompt, your computer will display the following error message:

```
C:\TCLITE\BIN> Demo   <ENTER>
This program can only be run
from within the IDE
```

As you can see, the *Demo* program did not display its output message, but rather, a message that you must run the program from within the *Turbo C++ Lite* integrated development environment (IDE). If you want to create a program that you can give to another user to run, you must buy a different compiler, such as Borland C++ or Microsoft *Visual C++*.

SAVING YOUR C++ PROGRAM STATEMENTS TO A SOURCE FILE

In Lesson 2, you will use the *Turbo C++ Lite* editor to type in a C++ program. After you type the program statements, you will save your program to a file on disk. To save your program statements to a file, perform the following steps:

1. Select the File menu Save As option. *Turbo C++ Lite* will display the Save File As dialog box, as shown in Figure 1.8.

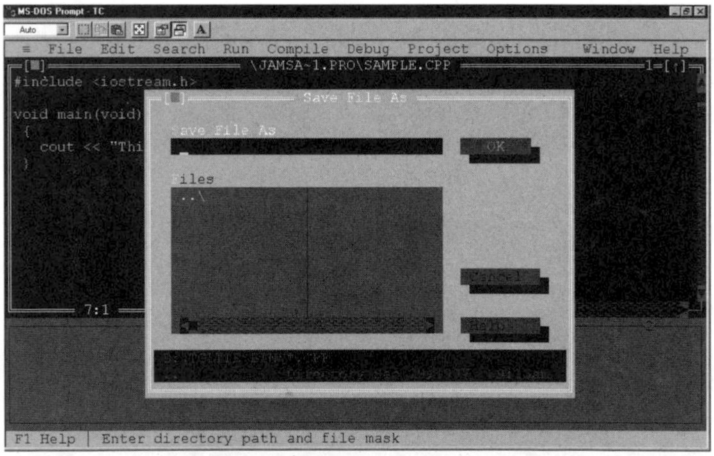

Figure 1.8 The Save File As dialog box.

2. Within the Save File As dialog box, type the filename that you desire.
3. Select OK.

To simplify your efforts, the CD-ROM that accompanies this book contains each of the programs this book's lessons present. Using the File menu Open option, you can open files that reside on the CD-ROM. If you make changes to the program code, you may want to save your statements into a file that resides on your hard disk. To save the file, select the File menu Save As option. Within the Save File As dialog box, select the drive, directory, and filename into which you want to save the file.

WHAT YOU MUST KNOW

In this lesson, you learned how to install and run the *Turbo C++ Lite* compiler. In Lesson 2, "Creating Your First C++ Program," you will create several small C++ programs. Using the *Turbo C++ Lite* compiler, you can type, compile, and run your programs. Before you continue with Lesson 2, however, make sure you have learned the following key concepts:

☑ The CD-ROM that accompanies this book contains the *Turbo C++ Lite* compiler, which is an MS-DOS-based compiler.

☑ Within *Turbo C++ Lite*, you can type C++ program statements, compile the program statements into the ones and zeros the computer can execute, and then run the program.

☑ Using the *Turbo C++ Lite* File menu Save option, you can save the C++ program statements to a source file on your disk.

☑ When you create executable program using *Turbo C++ Lite*, you can run the programs only from within the *Turbo C++ Lite* environment.

☑ To create an executable program that you can run on any computer, you must compile the program with such compilers as Borland C++ or Microsoft *Visual C++*.

LESSON 2

CREATING YOUR FIRST PROGRAM

Each of us has used computer programs such as a word processor, spreadsheet, Web browser, and even Microsoft Windows®. Computer programs, or *software*, are files that contain instructions that tell the computer how to perform specific tasks. For example, a word-processing program contains instructions that tell the computer how to save, print, and even spell check your documents.

If you are working in the Windows environments, for example, files with the *EXE* and *COM* extensions contain executable commands the computer can perform. In other words, these program files contain specific instructions the computer performs, normally one after another, to accomplish a specific task. When you create programs, you use a programming language, such as C++, to specify the instructions you want the computer to perform. Then, using a special program called a compiler, you convert your C++ programming statements into the ones and zeros the computer understands.

In this lesson, you will learn how to specify computer instructions using C++ statements. By the time you finish this lesson, you will understand the following key concepts:

- To create a program, you use a text editor to type C++ statements into a program *source file*.

- To convert your C++ program statements into an executable program (the ones and zeros the computer understands), you use a special program called the C++ *compiler*.

- To make changes or corrections to your program, you use a text editor.

- When you violate one (or more) of the C++ programming rules, the C++ compiler will display syntax-error messages to your screen. When the compiler displays syntax errors, you must edit your program to correct the errors.

- Each time you change your C++ source file, you must compile your program to put your changes into effect.

Programming is the process of defining the list of instructions the computer must perform to accomplish a specific task. To specify the instructions, you use a *programming language,* such as C++. Using a text editor, you place the program statements in a *source file*. Next, you use a special program, *a compiler*, to convert the statements from a format that you can read and understand into the ones and zeros the computer understands.

The best way to understand the process of creating and compiling a program is to build a simple C++ program—which you will do right now!

CREATING A SIMPLE PROGRAM

Your first C++ program is named, as you might expect, *First.CPP*. As you create C++ programs, use the *CPP* extension to indicate to others (programmers mostly) that your file contains a C++ program. When you later execute the *First.CPP* program, it will display the message *Rescued by C++!* on your screen display. The following sample output shows a command line prompt (C:\> is shown in this example), the command line you type (the program name *First* followed by ENTER), and the program's screen output:

```
C:\> First   <ENTER>
Rescued by C++!
```

When you create programs, you can work in a command-line-based environment, such as MS-DOS or UNIX, or you can work in a Windows-based environment. For simplicity, this book's sample program output assumes you are working from a command line. In this case, to execute the program *First.EXE*, you would type the program name *First* at the system prompt and then press ENTER.

To create your program, you will use a text editor, such as EDIT (provided with MS-DOS) or the *Turbo C++ Lite* editor which is built into the Borland *Turbo C++ Lite* compiler included on this book's companion CD-ROM, to create the file (called a source file) that contains the program statements. Do not use a word processor such as *Word*® or *WordPerfect*® to create your program source file. As you know, word processors let you create formatted documents, with bold text, aligned margins, and other features. To format your documents in this way, the word processor embeds special (hidden) characters within your document. Such characters might turn italics on or off or select a specific margin width. Although these special characters are meaningful to your word processor, C++ will not understand them, and the characters will cause errors.

Using your text editor, type in the following C++ program statements (exactly as the characters appear, using upper and lowercase letters as shown):

```cpp
#include <iostream.h>

void main(void)
 {
    cout << "Rescued by C++!";
 }
```

Don't worry if the C++ statements do not make sense to you. You will learn each statement's purpose in Lesson 3 "Taking a Closer Look at C++." For now, you should pay particular attention to your typing. Make sure, for example, that you include the correct number of quotes, semicolons, and parentheses. Examine your program statements closely one more time. If they are correct, save the statements to the file *First.CPP*.

Creating and Running First.CPP within TCLITE

As you learned in Lesson 1, "Installing and Using Borland *Turbo C++ Lite*" the companion CD-ROM that accompanies this book includes the Borland *Turbo C++ Lite* compiler. To create the *First.CPP* program within *Turbo C++ Lite*, perform these steps:

1. Start *Turbo C++ Lite* as discussed in Lesson 1.
2. Select the File menu New option.
3. Within the *Turbo C++ Lite* code window, type in the *First.CPP* program statements as shown in Figure 2.1.

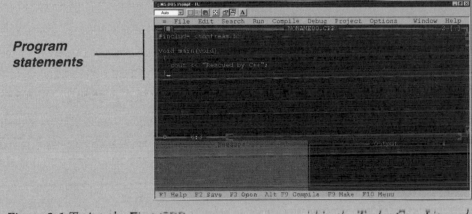

Figure 2.1 Typing the First.CPP program statements within the Turbo C++ Lite code window.

4. Select the File menu Save As option. *Turbo C++ Lite* will display the Save As dialog box. Type in the filename *First.CPP* and press ENTER.

5. Select the Run menu Run option. *Turbo C++ Lite* will compile the program statements and will then run the program, displaying its output within the *Turbo C++ Lite* output window as shown in Figure 2.2.

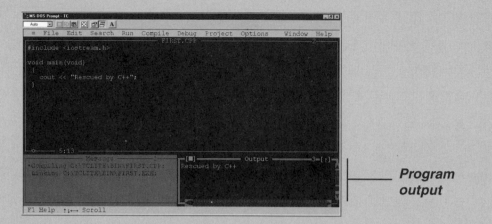

Program output

Figure 2.2 Displaying the output of First.CPP within the Turbo C++ Lite output window.

Assign Meaningful Names to Program Source Files

When you create C++ programs, you will place your program statements in a source file. When you name your program's source file, use the *CPP* extension to indicate to other programmers (and to the C++ compiler) that the file contains a C++ program. Next, use a filename that specifies the program's purpose. If you create a budget program, for example, you might use the name *Budget.CPP*. Likewise, for a program that calculates a company's salaries, you might call the program *Salary.CPP*. To avoid confusion, never use the name of an existing operating-system command, such as *COPY* or *DEL*, for a program name.

COMPILING YOUR PROGRAM

Computers work with combinations of ones and zeros (called *machine language*), which represent the presence or absence of electrical signals within the computer. If a signal is one (present), the computer may perform one operation, whereas if the signal is zero (absent), the computer may perform a different operation. Luckily, you don't have to write your programs in terms of ones and zeros (as did programmers in the 1940s and '50s). Instead, you can write your programs using a programming language, such as C++, and then use a special program, the C++ compiler, to convert your program statements (your source code) into machine language.

In other words, the compiler examines the source file that contains your C++ program statements. If your statements do not violate any of the C++ programming language rules, the compiler will convert the statements into machine language (ones and zeros) that the computer can execute. The compiler stores the machine language in an executable file, which typically uses the *EXE* extension. After the *EXE* file exists, you can run your program by typing its name at the command prompt and pressing ENTER.

Depending on the compiler you are using, the command you execute to use the compiler will differ. Refer to the documentation that accompanied your compiler to determine the correct command to start your compiler. After the compiler completes your program's source code, the compiler will create an executable program that it stores in a file **12** on your disk. If you are working in the MS-DOS environment, the executable program file will have the *EXE* extension, such as *First.EXE*.

Download a C++ Compiler

As you learned in Chapter 1, "Installing and Using the *Turbo C++ Lite* Compiler," The companion disk that accompanies this CD-ROM includes the Borland *Turbo C++ Lite* compiler, which runs under MS-DOS. If you are programming in the Unix or Mac environments, you may be able to download from the World Wide Web a shareware C++ compiler, which you can use to compile the programs this book presents. For more information on downloading C++ compilers, visit the Web site *http://www.ncf.carleton.ca/~bg283/* or *http://www.cs.princeton.edu/software/lcc/index.html.*

If your compiler displays error messages when you compile the program, edit your source file and compare each character in your file to the characters that appear in the book. Correct any mistakes, save your changes, and then compile the program a second time. After you successfully compile your program, execute the program by typing the program name at the command prompt as previously shown.

Understanding the Compiler

When you create a program, you use a programming language (such as C++) to specify the instructions the computer must execute to perform a specific task. Using a text editor, you place your program statements in a source file. Next, you use a special program called a compiler, which converts your source file to machine language (ones and zeros the computer understands). If the compiler is successful, it will create an executable program file. However, if you mistype one or more lines or if you violate any of the rules of C++, the compiler will display error messages on your screen, and you must edit your source file to correct the mistakes.

If you are working on a mainframe or mini-computer, you may have a compiler available that you and all other users on the system can access. If you are using a PC, you must buy and install a compiler such as Borland *C++* or Microsoft *Visual C++*.

CREATING A SECOND PROGRAM

You have successfully compiled and executed the *First.CPP* program. Now, use your text editor to create a second program file, named *Easy.CPP*, which contains the following program statements:

```
#include <iostream.h>

void main(void)
  {
    cout << "Programming in C++ is easy!";
  }
```

As before, save your C++ program statements to the source file and then use your compiler to create the executable program.

If the program successfully compiles, it will create an executable program named *Easy.EXE*. When you execute the program, your screen will display the following message:

```
C:\> Easy  <ENTER>
Programming in C++ is easy!
```

Next, using your editor, edit the source file *Easy.CPP* and change the program message to include the word *very*, as shown here:

```
    cout << "Programming C++ is very easy!";
```

Save your change to the source file and compile the program. After the compiler successfully completes, execute the program as shown here:

```
C:\> Easy   <ENTER>
Programming in C++ is very easy!
```

AFTER YOU CHANGE A SOURCE FILE, YOU MUST RECOMPILE

Each time you make changes to a program's source file, you must compile the program for the changes to take effect. For example, use your text editor to change the source file *Easy.CPP* a second time. This time, add the new statement in your program, as shown here:

```
#include <iostream.h>

void main(void)
 {
    cout << "Programming in C++ is easy!";
    cout << endl << "And pretty cool!";
 }
```

Save your changes to the source file. Next, execute the program as shown here:

```
C:\> Easy   <ENTER>
Programming in C++ is very easy!
```

As you can see, the program did not display your second line of output. Before your changes to a source file take effect, you must compile your program. Now compile the program as previously discussed and then execute it. Because the compiler has applied your source-code changes, your screen will display the second line of output as shown here:

```
C:\> Easy   <ENTER>
Programming in C++ is very easy!
And pretty cool!
```

UNDERSTANDING SYNTAX ERRORS

Every language, such as English, French, German, and even C++, has a set of syntax rules that you must follow when you use the language. In English, for example, sentences usually end with periods, exclamation points, or question marks. Likewise, you normally use capital letters to start sentences. In C++ the syntax uses semicolons, parentheses, braces, and many other characters. When you forget or misuse one of these characters, the C++ compiler will display an error message on your screen that describes the error. In addition, most compilers will display the errant statement's line number within your source file.

The C++ compiler cannot create an executable program until you correct all syntax errors. To understand the process of detecting and correcting syntax errors, create the following program, named *Syntax.CPP*:

```
#include <iostream.h>

void main(void)
  {
    cout << Use quotes around messages;
  }
```

If you look closely, you might notice that the messages the previous two programs displayed appeared in your source file within quotes. The C++ syntax (language rules) requires the quotes. When you compile the *Syntax.CPP* program, your compiler will display syntax error messages. In the case of the Borland *Turbo C++ Lite* compiler provided on this book's companion CD-ROM, the compiler will display the following messages:

```
Compiling Syntax.CPP:
Error Syntax.CPP 5: Undefined symbol 'Use' in function main()
Error Syntax.CPP 5: Undefined symbol 'quotes' in function main()
Error Syntax.CPP 5: Undefined symbol 'around' in function main()
Error Syntax.CPP 5: Undefined symbol 'messages' in function main()
```

In this example, the compiler displays four syntax errors. As you can see, all the errors relate to line 5 of the source file. Edit the file and place the quotes around the message, as shown here:

```
    cout << "Use quotes around messages";
```

You can now successfully compile the program to produce an executable file. When you first start using any programming language, you can count on having to correct several syntax errors each time you compile a program. After you have a few programs "under your belt," you will detect and correct such errors quickly.

Understanding Syntax Errors

When you create C++ programs, you must obey certain syntax rules. For example, you must place text messages within quotes and place a semicolon at the end of most statements within your program (you will learn later in this book which statements require semicolons and which do not). When your program breaks a syntax rule, the C++ compiler will display an error message on your screen. You must correct all syntax errors before the compiler can produce an executable program.

WORKING IN A WINDOWS-BASED ENVIRONMENT

For simplicity, each of the previous examples have assumed you are working in a command-line-based environment such as MS-DOS or UNIX. Most C++ programmers today, however, program in a Windows-based environment such as *Visual C++* or Borland's *C++ 5.02* for Windows. When you program in a Windows-based environment, your program statements do not change from those shown in this book's examples. In other words, the C++ statements in the program *First.CPP* are identical in a Windows-based environment to those you would use in a command-line-based environment. What does change in a Windows-based environment is how you compile and execute your programs.

Within a Windows-based programming environment, you can create your source files using a built-in editor and then compile the program by selecting a menu option or by clicking your mouse on a toolbar button. If the program contains syntax errors, the programming environment may display the error messages in a different window. After you successfully compile your program, you can use a menu option (or toolbar button) to execute the program. **15**

Again, the programming environment may open a separate window within which it will display the program's output. Figure 2.3 illustrates a Windows-based programming environment.

Figure 2.3 A Windows-based programming environment.

A programming environment is so named because it provides all the tools you must have to create, compile, and execute programs.

WHAT YOU MUST KNOW

In this lesson you learned how to create and compile C++ programs. In Lesson 3, "Taking a Closer Look at C++," you will take a closer look at the statements that make up the programs you created in this lesson. You will learn the purpose of keywords such as *void,* how the programs know to display output on the screen, and that C++ programs use braces {} to group related statements. Before you continue with Lesson 3, however, make sure you understand the following key concepts:

☑ Programs are files that contain a list of instructions the computer will execute to perform a specific task.

☑ To create C++ programs you will use a text editor to type in your program statements.

☑ You will store your C++ programs in source files that use the *CPP* extension.

☑ The compiler converts the C++ program statements into 1s and 0s (machine language) that the computer understands.

☑ Like all languages, C++ has a set of language rules, or syntax.

☑ If you break a C++ syntax rule, the compiler will display a message describing the error and possibly the line number in your program that is causing the error.

☑ You must remove all syntax errors before the compiler will create an executable program.

☑ After you make changes to your source file, you must compile your program for the changes to take effect.

LESSON 3

TAKING A CLOSER LOOK AT C++

In Lesson 2 "Creating Your First Program," you created several C++ programs. At that time, your goal was not to understand the C++ statements, but rather, to understand the process of creating and compiling C++ programs. In this lesson, you will take your first close look at the statements that make up a C++ program. You will find that most C++ programs follow the same format, beginning with one or more *#include* statements, a line consisting of *void main(void)* and then a set of statements the program groups using right and left braces {}. In this lesson, you will learn that these somewhat intimidating statements are actually very easy to understand. By the time you finish this lesson, you understand the following key concepts:

- The *#include* statement lets you take advantage of *header files* that contain C++ statements or program definitions.

- The main part of a C++ program begins with the statement *void main(void)*.

- Programs consist of one or more functions that contain related statements that accomplish a specific task.

- To display screen output, your programs will make extensive use of the *cout* output stream.

When you create C++ programs, you actually work in terms of *statements*, not instructions. In later lessons you will learn about the *assignment statement*, which assigns values to variables, the *if* statement, which lets your program make decisions, and so on. For now, we will simply refer to the contents of your program as *program statements*.

LOOKING AT PROGRAM STATEMENTS

In Lesson 2, you created the C++ program, *First.CPP*, which contained the following statements:

```
#include <iostream.h>

void main(void)
  {
    cout << "Rescued by C++!";
  }
```

In this example, the program contains three statements. The braces (called *grouping symbols*) group related statements, as shown here:

```
#include <iostream.h>

void main(void)                              Program Statements
  {
    cout << "Rescued by C++!";
  }
```

The following sections look at each of the program's statements in detail.

UNDERSTANDING THE #INCLUDE STATEMENT

Each of the programs you created in Lesson 2 began with the following *#include* statement:

```
#include <iostream.h>
```

When you create C++ programs, you can take advantage of statements and definitions that the compiler provides for you. When you compile your program, the *#include* statement will direct the compiler to include, at the current location within your program's source file, the contents of the header file you specify between the brackets. In this case, the compiler will include the contents of the header file *iostream.h*.

Files with the *.h* extension that you include at the start (or *head*) of your programs are *header files*. If you examine the directory that contains your compiler files, you should find a subdirectory named *INCLUDE*, which contains many different header files. Each header file contains definitions the compiler provides for different operations. For example, there is a header file that provides definitions for math operations, a different header file for file operations, and so on.

The header files are ASCII files, so you can display their contents on the screen or print them. For now, don't worry about the header file contents. Instead, simply understand that the *#include* statement lets you use these files. Each of the C++ programs you will create throughout this book will specify one or more *#include* statements.

Taking a Look at C++ Header Files

 Each C++ program you create will begin with one or more *#include* statements. The *#include* statements direct the compiler to insert (include) a specific file's contents (a header file) within your program, just as if your program contained the statements the file contains. Header files contain definitions the compiler uses for different types of operations. There are header files that define C++ input/output (I/O) operations, operating-system services (such as functions that return the current date and time), and much more.

Header files, like your C++ programs, are ASCII files whose contents you can view or print. To better understand the header file's contents, take time now to print the header file *iostream.h*, whose contents you will use in each C++ program you create throughout this book. You will normally find the *iostream.h* header file within a subdirectory named *INCLUDE*, which resides in the directory that contains your C++ compiler files. Use your text editor to view and print the header file's contents.

Note: Never change a header file's contents. Doing so may cause your compiler to experience errors in each program you create.

UNDERSTANDING VOID MAIN(VOID)

When you create a C++ program, your source file will contain many statements. As you will learn, the order in which the statements appear in your program is not necessarily the order in which the computer will execute the statements when it runs the program. Every C++ program has one location at which the program's execution begins—the *main program*. In C++ programs, the *void main(void)* statement specifies your program's starting location.

As your programs become larger and more complex, you will divide them into smaller, more manageable pieces. At that time, the *void main(void)* statement will identify your program's primary (or main) statements—the part of the program that the computer executes first.

Understanding the Main Program

C++ source files may contain many statements. When you run a program, the statement *void main(void)* will identify your main program, which contains the first statement the computer will execute. Your C++ programs must always have one—and only one—statement that includes the name *main*.

As you examine larger C++ programs, search for the statement that contains the name *main* to determine the statements at which the computer will begin the program's execution.

UNDERSTANDING THE USE OF VOID

As your programs become more complex, you will break the program into smaller more manageable pieces called *functions*. A function is simply a set of statements within your program that perform a specific task. For example, if you create a payroll program, you might create a function you name *salary* that calculates an employee's salary. Likewise, if you write a math program, you might create functions you name *square_root* or *cube* that return the result of specific mathematical operations. When your program uses a function, the function will perform its task and then return its result to the program.

For now, each function within your program should have a unique name. Also, each program has at least one function. In the programs you created in Lesson 2, each program had only one function, *main*. Lesson 10 "Getting Started with Functions," takes a closer look at a functions. For now, simply understand that a function consists of one more related statements that accomplish a specific task.

As you examine different C++ programs, you will regularly encounter the word *void*. Programs use the word *void* to specify that a function does not return a value or that the program does not pass parameters to the function.

Most of the simple C++ programs you will create throughout this book will not return an exit-status value to the operating system. Therefore, you will place word *void* in front of *main*, as shown here:

```
void main(void)              Program does not return a value
```

In later lessons you will learn that your programs can use information (such as a filename) the user specifies in the command line when he or she executes the program. When a program does not use command-line information, you place the word *void* between the parentheses that follow *main*, as shown here:

```
void main(void)              Program does not use
                             command-line arguments
```

As your programs become more complex, they might eventually return values to the operating system or support command-line parameters. For now, however, simply precede *main* with the *void* keyword.

Using a Program's Exit Status Value

Depending on your program's processing, there may be times when you run your program from within a batch file. Depending on the program's success, you may or may not want the batch file to run other programs. Using the *exit* function, your C++ programs can return a value to the operating system when they end. A batch file, in turn, can then test the program's exit value and continue its processing accordingly. For example, under MS-DOS, batch files test a program's exit-status value using the *IF ERRORLEVEL* command. Assume, for example, that a program named *Payroll.EXE* exits with one of the following exit status values, depending on the program's processing success:

Status Value	Meaning
0	Successful
1	File not found
2	Printer out of paper

Within a MS-DOS batch file, you can test the program's success using the *IF ERRORLEVEL* command, as shown here:

```
PAYROLL
IF ERRORLEVEL 0 IF NOT ERRORLEVEL 1 GOTO SUCCESSFUL
IF ERRORLEVEL 1 IF NOT ERRORLEVEL 2 GOTO NO_FILE
IF ERRORLEVEL 2 IF NOT ERRORLEVEL 3 GOTO NO_PAPER
REM Other batch file commands here
```

UNDERSTANDING GROUPING STATEMENTS {}

As your programs become more complex, you will have one set of statements that you might want the computer to execute a specific number of times and a different set of statements you want the computer to execute only if a specific condition is true. In the first case, the computer might perform the same set of statements 100 times to add up 100 student test scores. In the second case, the computer might display one message to the screen if all the students passed the test and a different message if one or more students failed. Within your C++ programs you will use the left and right braces {} to group related statements. In the simple programs this book's first few lessons present, the symbols group the statements that correspond to your main program.

USING COUT TO DISPLAY SCREEN OUTPUT

Each of the C++ programs you created in Lesson 2 displayed a message to your screen. To display the message, the programs used *cout* and the double less-than sign (<<), as shown here:

```
cout << "Hello, C++!";
```

The word *cout* is an *output stream* that C++ associates with the operating system's standard-output device. By default, the operating system associates the standard-output device with the screen display. To display a message to the screen display, you simply use the double less-than symbol (called the insertion operator) with the *cout* output stream. In Lesson 4, "Writing Messages to Your Screen," you will learn that you can use the insertion operator to send characters, numbers, and other symbols to the screen display.

Understanding the cout Output Stream

As you have learned, C++ programs use the *cout* output stream to display messages to the screen. When you use *cout* to display messages, think of *cout* as aligning a stream of characters the operating system will display on the screen. In other words, the order your program sends characters to *cout* defines the order the characters will appear on the screen. As an example, consider the following program statements:

```
cout << "This message appears first, ";
cout << "followed by this message.";
```

In this case, the program will display the stream of characters as follows:

```
This message appears first, followed by this message.
```

The insertion operator (<<) is so named because the operator lets your program insert characters into the output stream.

As you have learned, the *cout* output stream, by default, corresponds to your screen display. In other words, when your programs send output to *cout*, the output appears on the screen. However, using the operating system's output redirection operators, you can send a program's output to the printer or a file. For example, the following command directs MS-DOS to send the output of the program *First.EXE* to the printer, as opposed to the screen display:

```
C:\> First > PRN   <ENTER>
```

As you will learn in Lesson 4, C++ lets you use *cout* to output characters, whole numbers, such as 1001, and floating-point numbers, such as 3.12345. In Lesson 7, "Reading Keyboard Input," you will learn that C++ also provides an input stream, named *cin*, that your programs can use to read information a user types at the keyboard.

WHAT YOU MUST KNOW

This lesson examined several common items you will encounter in C++ programs. In Lesson 4, "Writing Messages to Your Screen" you will learn how to use *cout* to display characters, whole numbers and floating-point values. You will also learn how to format such output. Before you continue with Lesson 4, however, make sure you have learned the following key concepts:

- ☑ Most C++ programs begin with the *#include* statement, which directs the compiler to include the contents of the specified header file within the program.

- ☑ Header files contain compiler definitions your programs can use.

- ☑ A source file may consist of many statements. The statement *void main(void)*, specifies the start of the main program, which contains the first statement the program will execute.

- ☑ The first *void* in the statement *void main(void)* tells the compiler (and another programmer who is reading your code) that the program does not return a value to the operating system.

- ☑ The second *void* in the statement *void main(void)* tells the C++ compiler (and programmers who are reading your code) that the program does not support command-line parameters.

- ☑ As your programs become more complex, you will group related statements within smaller, more manageable pieces called functions.

- ☑ To group related program statements, you will use right and left braces {}.

- ☑ Most C++ programs use the *cout* output stream to display information to the screen. Using the operating system's I/O redirection operators, you can redirect *cout's* output to a file, a device (such as the printer), or even to become the input of a second program.

LESSON 4

WRITING MESSAGES TO YOUR SCREEN

Each of the C++ programs you created in Lesson 2, "Creating Your First Program," and Lesson 3, "Taking a Closer Look at C++," used the *cout* output stream to display messages (the program's output) to the screen. In this lesson you will use *cout* to display characters, whole numbers such as 1001, and floating-point numbers such as 0.12345. By the time you finish this lesson, you will understand the following key concepts:

- C++ programs use the *cout* output stream to display characters and numbers to the screen.

- C++ lets you use special characters with *cout* to generate a tab or newline and even to sound your computer's built-in speaker.

- C++ makes it easy for you to display numbers in decimal, octal (base 8), or hexadecimal (base 16) format.

- Using operating-system command-line redirection operators, you can redirect output messages your programs send to *cout* from the screen display to a file, to a printer, and even to a second program.

- Using the *cerr* output stream, your programs can send messages to the standard-error device, which prevents users from redirecting the message from the screen.

- Using the *setw* modifier within an output stream, your programs can format their output.

Almost every C++ program you create will use *cout* to display screen messages. This lesson will teach you how to make the best use of *cout.*

USING COUT TO DISPLAY NUMBERS

In this book's previous lessons, your programs used *cout* to display *character strings* (letters and numbers contained within quotes) to the screen. As you will learn in this section, you can also use *cout* to display numbers. The following program, *1001.CPP*, displays the number 1001 on your screen display:

```cpp
#include <iostream.h>

void main(void)
  {
    cout << 1001;
  }
```

After you compile and execute this program, your screen will display the number 1001, as shown here:

```
C:\> 1001   <ENTER>
1001
```

Next, edit the program and change the *cout* statement to display the number 2002, as shown here:

```cpp
cout << 2002;
```

Remember, after you save your edits to the source file, you must recompile the program for your changes to take effect.

In addition to letting your programs display *whole numbers* (numbers without decimal points), *cout* also lets your programs display *floating-point numbers*, such as 1.2345. The following program, *Floating.CPP*, uses *cout* to display the number 0.12345 on your screen:

```
#include <iostream.h>


void main(void)
 {
    cout << 0.12345;
 }
```

As before, compile and execute the *Floating.CPP* program. Your screen, in turn, will display the following output:

```
C:\> Floating   <ENTER>
0.12345
```

DISPLAYING SEVERAL VALUES AT ONE TIME

C++ programmers refer to the double less-than sign as the *insertion operator* (the operator inserts characters into the output stream for display). When you use *cout*, you can use the insertion operator more than one time per statement. For example, the following program, *1001Too.CPP*, uses the insertion operator four times to display the numbers 1001 on your screen:

```
#include <iostream.h>

void main(void)
 {
    cout << 1 << 0 << 0 << 1;
 }
```

After you compile and execute the *1001Too.CPP* program, your screen will display the following:

```
C:\> 1001Too   <ENTER>
1001
```

Each time C++ encounters the insertion operator, it simply appends the number or characters that follow to those currently in the output stream. The following program, *Show1001.CPP*, displays a character string and a number using *cout*:

```
#include <iostream.h>

void main(void)
 {
   cout << "My favorite number is " << 1001;
 }
```

Note that you use the space character that follows the word *is* (within quotes) to place the number 1001 one space from the word. Without the space, the number would appear next to the word (is1001). In a similar way, the following program, *1001Mid.CPP*, displays the number 1001 in the middle of a character string:

```
#include <iostream.h>

void main(void)
  {
    cout << "The number " << 1001 << " is my favorite";
  }
```

As before, note the spacing in the character strings that appear before and after 1001. If you do not include the space characters after the word number and before the word *is*, the program will place the number 1001 immediately next to the words (number1001is).

Finally, the following program, *MixMatch.CPP*, combines strings, characters, whole numbers, and floating point numbers within the same output stream:

```
#include <iostream.h>

void main(void)
  {
    cout << "At age " << 20 << " my salary was " << 493.34 << endl;
  }
```

After you compile and execute the *MixMatch.CPP* program, your screen will display the following output:

```
C:\> MixMatch    <ENTER>
At age 20 my salary was 493.34
```

USING SPECIAL OUTPUT CHARACTERS

All the programs you have created so far in this book displayed their output on one line. Most of the programs you will create in the future, however, will display several lines of output. For example, assume that you are writing a program that will display your address on the screen. Presumably, you will want your address to appear on multiple lines—just as it would appear on an envelope.

When you want to move the cursor to the start of the next line, you can place the *newline character* (\n) in the output stream. C++ gives you two different ways to generate the newline character. First, you can place the character \n within a character string. For example, the following program, *TwoLines.CPP*, includes the newline character within the output string, which causes the program to display its output on two lines:

```
#include <iostream.h>

void main(void)
  {
    cout << "This is line one\nThis is line two";
  }
```

When you compile and execute the *TwoLines.CPP* program, the newline character will cause the program to display two lines of output, as shown here:

```
C:\> TwoLines    <ENTER>
This is line one
This is line two
```

If your program is not displaying a character string, you can place the newline character within single quotes. For example, the following program, *NewLines.CPP*, displays the numbers 1, 0, 0, and 1, each on its own line:

```
#include <iostream.h>

void main(void)
  {
    cout << 1 << '\n' << 0 << '\n' << 0 << '\n' << 1;
  }
```

C++ treats the newline symbol as a character, much like the letters 'a', 'b', or 'c'. As such, when you use the newline character outside of a character string, you must enclose the newline character within single quotes. In addition to using the newline character to advance the cursor to the start of the next line, your programs can use the *endl* (end line) symbol. The following program, *Endl.CPP*, illustrates how to use *endl* to advance the cursor to the start of a new line:

```
#include <iostream.h>

void main(void)
  {
    cout << "I've been..." << endl << "Rescued by C++";
  }
```

As before, after you compile and execute the *Endl.CPP* program, your screen will display the program's output on two lines:

```
C:\> Endl     <ENTER>
I've been
Rescued by C++
```

Finally, the following program, *Address.CPP*, displays the Jamsa Press address on multiple lines:

```
#include <iostream.h>

void main(void)
  {
    cout << "Jamsa Press" << endl;
    cout << "2975 South Rainbow, Suite I" << endl;
    cout << "Las Vegas, NV 89102" << endl;
  }
```

In this case, the *Address.CPP* program uses three program statements to output the address. As it turns out, however, the program could simply use one program statement that writes all the output to the screen as shown here:

```
#include <iostream.h>

void main(void)
  {
    cout << "Jamsa Press" << endl << "2975 South Rainbow, Suite I" << endl
       << "Las Vegas, NV 89102" << endl;
  }
```

25

As you can see, the program uses one statement to place the address into the *cout* output stream. Although two programs perform the same task, most programmers will find the first program, which uses three statements, easier to understand.

Other Special Characters

In addition to the newline character, which lets your programs advance the cursor to the start of a new line, you can use the special characters listed in Table 4 within the *cout* output stream.

Character	Purpose
\a	Alert (or bell) character
\b	Backspace character
\f	Formfeed character
\n	Newline character
\r	Carriage return (no linefeed)
\t	Horizontal tab character
\v	Vertical tab character
\\	Backslash character
\?	Question mark character
\'	Single quote character
\"	Double quote character
\0	Null character
\ooo	Octal value, such as \007
\xhhhh	Hexadecimal value, such as \xFFFF

*Table 4 Special characters your programs can use within the **cout** output stream.*

Note: When you use the special characters listed in Table 4, you must place the characters within single quotes when you use the characters by themselves, such as '\n'; or within double quotes if you use the characters within a string, such as "Hello\nWorld!"

The following program, *Special.CPP*, uses the alert (\a) and tab (\t) special characters to sound the computer's built-in speaker, and then display the words *Bell Bell Bell*, separated by one tab each:

```
#include <iostream.h>

void main(void)
  {
    cout << "Bell\a\tBell\a\tBell\a";
  }
```

Displaying Octal and Hexadecimal Values

So far, this lesson's programs have displayed numbers in decimal. Depending on your program's purpose, there may be times when you must display output in octal or hexadecimal. To do so, your programs can place the *dec, oct,* and *hex* output modifiers within your output stream. The following program, *OctHex.CPP*, uses these three modifiers to display the decimal values 10 and 20 in octal and hexadecimal:

```
#include <iostream.h>

void main(void)
 {
   cout << "Octal: " << oct << 10 << ' ' << 20 << endl;
   cout << "Hexadecimal: " << hex << 10 << ' ' << 20 << endl;
   cout << "Decimal: " << dec << 10 << ' ' << 20 << endl;
 }
```

After you compile and execute the *OctHex.CPP* program, your screen will display the following output:

```
C:\> OctHex   <ENTER>
Octal: 12 24
Hexadecimal: a 14
Decimal: 10 20
```

Note: *When you use one of the output modifiers to select octal, hexadecimal, or decimal output, your selection will remain in effect until the program ends or you use a different modifier.*

WRITING OUTPUT TO THE STANDARD ERROR DEVICE

As you have learned, when your programs use *cout* to display output, you can later redirect the program's output to a device or file using the operating system's output-redirection operators. When your programs encounter an error, however, you will not want the operating system to redirect error messages from the screen. For example, if the operating system redirects the error messages to a file, the user may be unaware that an error occurred.

When you want your programs to display an error message, you should use the *cerr* output stream. C++ associates *cerr* with the operating system's standard-error device. The following program, *Cerr.CPP*, uses the *cerr* output stream to display the message "*This Message Always Appears*" to the screen display:

```
#include <iostream.h>

void main(void)
 {
   cerr << "This Message Always Appears";
 }
```

Compile and execute the *Cerr.CPP* program. Next, try to redirect the program's output to a file using the output redirection operator, as shown here:

```
C:\> Cerr  >  FileName.EXT   <ENTER>
```

Because the operating system will not let your programs redirect output written to the standard error device, the message will appear on your screen display.

CONTROLLING THE OUTPUT WIDTH

Several of the previous programs used the *cout* output stream to display numbers to the screen. To ensure that the numbers display correctly (with the proper spacing), the programs included space characters before and after the numbers. When you use *cout* or *cerr* to display output, your programs can specify each value's output width using the *setw* (set width) modifier.

Using *setw*, your programs specify the minimum number of characters the value will consume. For example, the following program, *Setw.CPP*, uses the *setw* modifier to select a width of 3, 4, 5, and 6 for the value 1001. To use the *setw* modifier, you must include the header file *iomanip.h* at the start of your program, as shown here:

```
#include <iostream.h>
#include <iomanip.h>

void main(void)
 {
  cout << "My favorite number is" << setw(3) << 1001 << endl;
  cout << "My favorite number is" << setw(4) << 1001 << endl;
  cout << "My favorite number is" << setw(5) << 1001 << endl;
  cout << "My favorite number is" << setw(6) << 1001 << endl;
 }
```

After you compile and execute the *Setw.CPP* program, your screen will display the following output:

```
C:\> Setw   <ENTER>
My favorite number is1001
My favorite number is1001
My favorite number is 1001
My favorite number is  1001
```

When you specify a width using *setw*, you will specify the *minimum* number of characters the value will consume. In the example of the *Setw.CPP* program, the modifier *setw(3)* specified a minimum of three characters. However, because the value 1001 requires more than three characters, *cout* used the actual number it needs, which in this case is four. Also, when you use *setw* to select a width, the output stream will use the width only for the next value output. If you must specify a width for multiple values, you must use *setw* multiple times.

Note: The *Setw.CPP* program uses the header file **iomanip.h**. You may want to take time now to print and examine the file's contents. As was the case with the header file **iostream.h**, you should find the file in the **INCLUDE** subdirectory within the directory that contains your compiler files.

WHAT YOU MUST KNOW

In this lesson you learned how to use *cout* in many different ways to display screen output. Each of the programs you will create throughout the remainder of this book will use *cout* to display their output. In Lesson 5, "Programs Store Information in Variables," you will learn how to use variables within your programs to store values that may change as your program executes. Before you continue with Lesson 5, however, make sure that you have learned the following key concepts:

☑ The *cout* output stream lets you display characters and numbers to the operating system's standard-output device—which is normally the screen.

☑ Using special symbols within your output stream, your programs can specify newline, tab, and other special characters.

☑ To advance the cursor to the start of the next line, your programs can create a newline using the newline character \n or the *endl* modifier.

☑ The *dec, oct, and hex* modifiers let your programs display values in decimal, octal, and hexadecimal.

☑ Using the *cerr* output stream, your programs can write messages to the operating system's standard-error device—whose output the user cannot redirect.

☑ Using the *setw* modifier, your programs can control a value's output width.

LESSON 5

PROGRAMS STORE INFORMATION IN VARIABLES

All the programs presented in Lesson 2, "Creating Your First Program," through Lesson 4, "Writing Messages to Your Screen," have been very simple. As your programs begin to perform meaningful work, the programs must store information while they execute. For example, a program that prints a file must know the file's name and possibly the number of copies you want to print. As the program runs, it will store such information in your computer's random access memory (RAM). To store and retrieve information from specific memory locations, your programs use *variables*. In the simplest sense, a variable is the name of a memory location that can store a specific value. This lesson examines how to create and use variables within your C++ programs. By the time you finish this lesson, you will understand the following key concepts:

- You must declare the variables you will use within your program by telling the compiler the variable's name and type.

- A variable's *type* specifies the kind of value (such as a whole or floating-point number) the variable can store as well as the operations your program can perform on the variable.

- To assign a value to a variable, you use the C++ assignment operator (the equal sign).

- To display a variable's value to the screen, your programs use the *cout* output stream.

- When you declare variables, you should use meaningful names to make your programs easier for other programmers to read and understand.

- Within your program, you should place comments that describe your program's operations to other programmers. Therefore, if another programmer must change your program, your comments will describe your program's operations in detail.

Think of a variable as a box into which you can place a value. When you assign a value to the variable, you place a value within the box. When you later use the variable's value, the computer will simply look at the value the box contains.

DECLARING VARIABLES WITHIN YOUR PROGRAMS

Your programs use variables to store information. Depending on the kind of value you want to store, such as a whole number, letter of the alphabet, or floating-point value, your variable's *type* will differ. A variable's type specifies the kind of value the variable can store as well as the set of operations (such as addition, multiplication, and so on) the program can perform on the variable's value. Most C++ programs will use the variable types listed in Table 5.1.

Type	Values Stored
char	Values in the range −128 through 127. Programs normally use the type *char* to store letters of the alphabet
int	Values in the range −32,768 through 32,767
unsigned	Values in the range 0 through 65,535
long	Values in the range −2,147,483,648 through 2,147,483,647
float	Values in the range −3.4 x 10−38 through 3.4 x 1038
double	Values in the range 1.7 x 10−308 through 1.7 x 10308

Table 5.1 Common C++ variable types.

Before your program can use a variable, your program must *declare* the variable. In other words, your program must introduce the variable to the C++ compiler. To declare a variable in your program, you must specify the variable's type and the name your program will use to refer to the variable. For example, following the opening brace of your main program, you will specify the variable type and name, as shown here:

```
variable_type  variable_name;
```

Normally, the variable's type will be one of the types listed in Table 5.1. The variable name is a meaningful name you choose that describes (to someone reading your program) the variable's use. For example, your program might use variables such as *employee_name*, *employee_age*, and so on. Note the semicolon that follows the variable name within the variable's declaration. C++ considers a variable's declaration a statement. Therefore, you must place a semicolon at the end of the declaration.

The following program fragment declares three variables using the types *int*, *float*, and *long*:

```
#include <iostream.h>

void main(void)
 {
    int test_score;
    float salary;
    long distance_to_mars;
 }
```

It is important to note that this program does not do anything but declare the three variables. In other words, if you compile and run this code fragment, it will not display any output. As you can see, each variable declaration ends with a semicolon.

When you declare more than one variable of the same type, C++ lets you separate the variable names using a comma between names. The following statement, for example, declares three floating-point variables:

```
float salary, income_tax, retirement_fund;
```

Understanding Variables

A variable is the name of a storage location in your computer's random access memory (RAM). When a program runs, the program stores information in variables. When you create your programs, you must declare variables by telling the C++ compiler the variable's type and name. The following statement, for example, declares a variable of type *int* named *age*:

```
int age;
```

As you will learn, a variable's type specifies a set of values the variable can store (such as whole or floating-point numbers) and a set of operations the program can perform on the variable (such as addition and subtraction).

USE MEANINGFUL VARIABLE NAMES

Each variable you create in your program must have a unique name. To make your programs easier to read and understand you should use meaningful variable names. For example, the following statement declares three variables named *x*, *y*, and *z*:

```
int x, y, z;
```

Assuming the variables store a student's age, test score, and grade, the following variable names are much more meaningful to another programmer who is reading your source code:

```
int student_age, test_score, grade;
```

When you select variable names, you can use a combination of letters, numbers, and underscores (_). The first character of your variable names must be a letter or underscore. You cannot begin a variable name with a number. Also, C++ considers upper and lowercase letters different. As you get started, only use lowercase letters within your variable names. As you become more comfortable with C++, you might combine upper and lowercase letters to produce a meaningful name, as shown here:

```
float MonthlySalary, IncomeTax;
```

WORDS YOU CANNOT USE FOR VARIABLE NAMES

When you create variable names you must know that C++ reserves the words listed in Table 5.2 as keywords that have special meaning to the compiler. You cannot use a C++ keyword for a variable name.

C++ Keywords

asm	auto	break	case	catch	char
class	const	continue	default	delete	do
double	else	enum	extern	float	for
friend	goto	if	inline	int	long
new	operator	private	protected	public	register
return	short	signed	sizeof	static	struct
switch	template	this	throw	try	typedef
union	unsigned	virtual	void	volatile	while

Table 5.2 *C++ keywords.*

Why Your Programs Use Variables

As your programs become more complex, they may perform operations on many different items. For example, if you write a payroll program, the program must process information for each employee. In such a program, you might use variables you name *employee_name*, *employee_id*, *employee_salary*, and so on. As your program starts, it will assign information about the first employee to these variables. After your program calculates that employee's salary, the program will repeat the process for the next employee. To process the second employee's salary, the program will assign that employee's information (his or her name, ID, and salary) to the variables listed earlier, and then the program will perform its processing. In other words, as the program executes, it will assign different values to the variables—which in turn, will change or "vary" the variable's value.

ASSIGNING A VALUE TO A VARIABLE

As you have read, variables store values as your programs execute. After you declare a variable, you use the C++ *assignment operator* (the equal sign) to assign a value to a variable. The following statements assign values to several different variables. Note that you use a semicolon to end each statement, as shown here:

```
age = 32;
salary = 25000.75;
distance_to_the_moon = 238857;
```

Note: *When you assign values to variables, do not include commas within the values (such as 25,000.75 and 238,857). If you include the commas, the C++ compiler will generate and display syntax error messages.*

The following program fragment declares the variables just described and then uses the assignment operator to assign values to the variables:

```
#include <iostream.h>

void main(void)
  {
      int age;
      float salary;
      long distance_to_the_moon;

      age = 32;
      salary = 25000.75;
      distance_to_the_moon = 238857;
  }
```

Again, it is important that you recognize that this program does not display any output. Instead, the program's purpose is to show you how to assign values to one or more variables.

ASSIGNING A VALUE AT DECLARATION

When you declare a variable, it is often convenient to assign the variable's starting value at the same time (programmers refer to this process as "initializing a variable"). To make it easy for you to initialize variables, C++ lets you assign values when you declare variables, as shown here:

```
int age = 32;
float salary = 25000.75;
long distance_to_the_moon = 238857;
```

Many of the programs this book presents will assign values to variables at declaration.

Assigning Values to a Variable

Variables store information during a program's execution. To store a value within a variable, your programs must use the C++ assignment operator (the equal sign). The following statement uses the assignment operator to assign the value 5 to the variable *lesson*:

```
lesson = 5;
```

To simplify the process of assigning values to variables, C++ also lets you assign a value to a variable when you declare the variable, as shown here:

```
int lesson = 5;
```

USING A VARIABLE'S VALUE

After you assign a value to a variable, your programs can use the variable's value simply by referring to the variable name. The following program, *ShowVars.CPP*, assigns values to three variables and then displays each variable's value using *cout*:

```cpp
#include <iostream.h>

void main(void)
  {
    int age = 32;
    float salary = 25000.75;
    long distance_to_the_moon = 238857;

    cout << "The employee is " << age << " years old" << endl;
    cout << "The employee makes $" << salary << endl;
    cout << "The moon is " << distance_to_the_moon <<
        " miles from the earth" << endl;
  }
```

Note: The last cout statement does not fit on one line. In this case, the program simply wraps the words to the next line. Because C++ uses the semicolon to indicate the end of a statement, you can wrap the line in this way. When you must wrap one line to the next, try not to wrap the line within a character string (within double quotes) and then indent the wrapped line one or two spaces, as shown.

When you compile and execute the *ShowVars.CPP* program, your screen will display the following output:

```
C:\> ShowVars   <ENTER>
The employee is 32 years old
The employee makes $25000.75
The moon is 238857 miles from the earth
```

As you can see, to use a variable's value, you simply refer to the variable's name within your program. Before you continue, take time to change the program's source code by assigning different values to the variables *age* and *salary*, as shown here:

```
int age = 44;
float salary = 52000.50;
```

Compile and execute your program. By changing the variable's values, you will change the program's output, as shown here:

```
C:\> ShowVars  <ENTER>
The employee is 44 years old
The employee makes $52000.50
The moon is 238857 miles from the earth
```

EXCEEDING A VARIABLE'S STORAGE CAPACITY

As you have learned, a variable's type defines the set of values the variable can store. For example, a variable of type *int* can store values in the range -32,768 through 32,767. When you assign a value to a variable that is outside of a type's value range, an *overflow* error will occur. For example, the following program, *Overflow.CPP*, illustrates how exceeding a variable's range of values will result in an error. The program will assign the value 40,000 to a variable of type *int*, which can only hold values up to 32,767. Next, the program will assign the value 40,000,000,000 to a variable of type *long*, which can hold values up to 2,147,483,647. Finally, the program will assign the value 210 to a variable of type *char*, which can only hold values up to 127, as shown here:

```
#include <iostream.h>

void main(void)
  {
    int positive = 40000;
    long big_positive = 4000000000;
    char little_positive = 210;

    cout << "positive now contains " << positive << endl;
    cout << "big_positive now contains " << big_positive << endl;
    cout << "little_positive now contains " << little_positive << endl;
  }
```

When you compile and execute the *Overflow.CPP* program, your screen will display the following output:

```
positive now contains -25536
big_positive now contains -294967296
little_positive now contains Ò
```

As you can see, the program assigns values to variables of type *int, long,* and *char* that exceed each type's storage range—which produces an overflow error. As you work with variables, you must keep in mind the range of values the variable's type can store. Overflow errors are subtle and can be difficult for you to detect and correct. Also, note the value the program displays for the variable *little_positive*. Because the variable is type *char*, the *cout* output stream tries to output the variable's value as character. In this case, the value shown corresponds to the extended-ASCII character for the value 210.

35

UNDERSTANDING PRECISION

In the previous section you learned that overflow errors occur when you assign a value to a variable that falls outside the range of values the variable's type can store. In a similar way, you also must understand that computers do not have unlimited *precision* (accuracy) with which they can store numbers. For example, when you work with floating-point numbers (values that have a decimal point), there are times when the computer cannot represent the number in its exact format. Such precision errors can be difficult for you to detect within your programs.

The following program, *Precise.CPP*, assigns a value less than 0.5 to variables of type *float* and *double*. Unfortunately, because the computer has a limited ability to represent numbers, the program's variable's do not actually contain the value the program assigns, but rather, the value 0.5, as shown here:

```cpp
#include <iostream.h>

void main(void)
 {
    float f_not_half = 0.49999990;
    double d_not_half = 0.49999990;

    cout << "Floating point 0.49999990 is " << f_not_half << endl;
    cout << "Double 0.49999990 is " << d_not_half << endl;
 }
```

When you compile and execute the *Precise.CPP* program, your screen will display the following output:

```
Floating point 0.49999990 is 0.5
Double 0.49999990 is 0.5
```

As you can see, the values the program assigns to the variables and the values the variables actually contain are not exactly the same. Such precision errors occur because the computer must represent numbers using a fixed number of ones and zeros. In many cases, the computer can represent numbers exactly. At other times, as shown in this program, the computer's representation of a number is close, but not exact. As you program, you must keep precision in the back of your mind. Depending on the values with which your programs are working, precision errors may arise which are very difficult for you to detect.

USE COMMENTS TO IMPROVE YOUR PROGRAM'S READABILITY

When your programs become more complex, the number of statements they contain may make the programs difficult to understand. Because other programmers may eventually have to understand, and possibly change, your programs, you must write your programs in the most readable manner possible. You can increase your program's readability by doing the following:

- Use meaningful variable names that describe the variable's use
- Maintain proper statement indentation and alignment (see Lesson 8)
- Use blank lines to separate unrelated statements
- Use comments that explain the program's processing

As you create programs, you can place notes within your source file that explain the program's processing. Such notes (programmers call the notes *comments*) not only help other programmers understand your program, they might also help you remember why your program uses specific statements, after you have not looked at the program for several months. To place a comment within your C++ programs, you simply place two forward slashes (//) within your program statements, as shown here:

```
// This is a comment
```

When the C++ compiler encounters the double slashes, the compiler ignores all text remaining on that line. At a minimum, you should place comments at the start of each program to specify who wrote the program, when, and why, as shown here:

```
// Program: Budget.CPP
// Programmer: Kris Jamsa
// Date Written: 1-10-98
//
// Purpose: Tracks monthly budget information.
```

As your program performs different processing, you should place comments in front of or next to specific statements that explain their purpose. For example, consider the following assignment statement:

```
distance_to_the_moon = 238857;    // Distance in miles
```

The comment to the right of the assignment statement provides additional information to someone reading your program. In the case of the previous statement, the comment tells another programmer who is reading your code that the value 238,857 represents the distance to the moon in miles.

New programmers often have difficulty determining when and what to comment. As a rule, *you cannot have too many comments in your programs.* However, make sure your comments are meaningful. The following comments provide no additional information to a programmer who is reading your code:

```
age = 32;              // Assign 32 to the variable age
salary = 25000.75;     // Assign 25000.75 to the variable salary
```

Your goal when using comments is to explain why specific processing occurs.

Adding Comments to Your Programs

As you create your programs, you should include comments that explain the program's processing. Should other programmers have to change your program, they can use your comments to understand your program's behavior. C++ programs normally use the double slashes to indicate a comment, as shown here:

```
// This is a C++ comment
```

When the C++ compiler encounters the double slashes, the compiler will ignore all text (remaining on the current line) that follows the slashes. Good programs should be easy to read and understand. Comments increase your program's readability.

Note: In addition to using comments to improve your program's readability, you should use blank lines to separate unrelated program statements. When the C++ compiler encounters a blank line, the compiler simply skips the line.

WHAT YOU MUST KNOW

In this lesson you learned that programs store information in variables as they execute. In short, a variable is a name your programs assign to a memory location within which the program stores information. Before your programs can use a variable, you must declare the variable's name and type. In Lesson 6, "Performing Simple Operations," you will learn how to perform simple operations, such as addition and subtraction, on variables. Before you continue with Lesson 6, however, make sure you have learned the following key concepts:

- ☑ To use variables within your programs, you must declare the variable's type and name.

- ☑ Variable names must be unique and should be meaningful to another programmer who reads your source code. A variable's name should correspond to the variable's purpose.

- ☑ Variable names must start with a letter or underscore.

- ☑ C++ considers upper and lowercase letters different.

- ☑ A variable's type determines the types of values a variable can hold. Common variable types include *char*, *int*, *float*, and *long*.

- ☑ Comments improve your program's readability by explaining the program's processing. C++ programs represent comments using double slashes (//).

LESSON 6

PERFORMING SIMPLE OPERATIONS

In Lesson 5, "Programs Store Information in Variables," you learned how to declare and use variables within your programs. As your programs become more complex, you will perform arithmetic operations such as addition, subtraction, multiplication, and division on the values your variables contain. This lesson examines how you use C++ arithmetic operators to perform these operations. By the time you finish this lesson, you will understand the following key concepts:

- To perform mathematical operations within your programs, you will use the C++ arithmetic operators.

- To ensure C++ evaluates arithmetic operations in a consistent manner, C++ assigns a precedence to each operator.

- Using parentheses within your arithmetic expressions, you can control the order in which C++ performs each operation.

- Many C++ programs add or subtract the value one from variables using the C++ increment (++) and decrement (--) operators.

After you learn to recognize the different C++ arithmetic operators, you will find that performing math operations is very easy!

BASIC MATH OPERATIONS

Regardless of your program's purpose, most C++ programs will add, subtract, multiply, or divide values. As you will learn, your programs can perform arithmetic operations on constants (such as 3 * 5) or on variables (such as *payment – total*). Table 6.1 lists the C++ basic math operators.

Operator	Purpose	Example
+	Addition	total = cost + tax;
–	Subtraction	change = payment – total;
*	Multiplication	tax = cost * tax_rate;
/	Division	average = total / count;

Table 6.1 The C++ basic math operators.

The following program, *ShowMath.CPP*, uses *cout* to display the result of several simple arithmetic operations:

```
#include <iostream.h>

void main(void)
  {
    cout << "5 + 7 = " << 5 + 7 << endl;
    cout << "12 - 7 = " << 12 - 7 << endl;
    cout << "1.2345 * 2 = " << 1.2345 * 2 << endl;
    cout << "15 / 3 = " << 15 / 3 << endl;
  }
```

39

Take a close look at the program statements. Note that each expression first appears within quotes, which causes the program to output the characters (such as 5 + 7 =) on your screen. Next, the program displays the result of the operation followed by a newline (*endl*). When you compile and execute this program, your screen will display the following output:

```
C:\> ShowMath   <ENTER>
5 + 7 = 12
12 - 7 = 5
1.2345 * 2 = 2.469
15 / 3 = 5
```

The *ShowMath* program performs arithmetic operations using constant values only. The following program, *MathVars.CPP*, performs arithmetic operations using variables:

```cpp
#include <iostream.h>

void main(void)
 {
   float cost = 15.50;         // The cost of an item
   float sales_tax = 0.06;     // Sales tax is 6 percent
   float amount_paid = 20.00;  // The amount the buyer paid
   float tax, change, total;   // Sales tax, buyer change and
                               // total bill


   tax = cost * sales_tax;
   total = cost + tax;
   change = amount_paid - total;

   cout << "Item Cost: $" << cost << "\tTax: $" << tax <<
     "\tTotal: $" << total << endl;

   cout << "Customer change: $" << change << endl;
 }
```

In this case, the program uses only floating-point variables. As you can see, the program assigns values to the variables at declaration. Next, the program performs arithmetic operations on the variables to determine the amount of sales tax, the total item cost, and the amount of customer change. When you compile and execute this program, your screen will display the following output:

```
C:\> MathVars   <ENTER>
Item Cost: $15.5         Tax: $0.93 Total: $16.43
Customer change: $3.57
```

Take time now to edit the program, changing the variable's values. You might, for example, assign the variable *cost* the value 25.00 and the variable *amount_paid* the value 100.00.

INCREMENTING A VARIABLE'S VALUE BY 1

As you program, a common operation you will perform is to add 1 to the value of an integer variable. For example, assume your program uses the variable named *count* to keep track of the number of files it has printed. Each time your program prints a file, your program will add 1 to *count's* current value. Using the C++ assignment operator, your program can increment *count's* value as shown here:

```
count = count + 1;
```

In this case, the program first obtains *count's* value and then adds 1 to that value. Next, the program stores the result of the addition back to the variable *count*. The following program, *IncCount.CPP*, uses the assignment operator to increment the variable *count* (which originally contains the value 1000) by 1 (assigning the result, 1001, to the variable):

```
#include <iostream.h>

void main(void)
 {
   int count = 1000;

   cout << "count's starting value is " << count << endl;
   count = count + 1;
   cout << "count's ending value is " << count << endl;
 }
```

When you compile and execute the *IncCount.CPP* program, your screen will display the following output:

```
C:\> IncCount   <ENTER>
count's starting value is 1000
count's ending value is 1001
```

Because incrementing a variable's value is a common operation within programs, C++ provides an *increment operator*, the double plus sign (++). The increment operator provides a shorthand way to add 1 to a variable's value. The following statements, for example, both increment the variable *count's* value by 1:

```
count = count + 1;                count++;
```

The following program, *Inc_Op.CPP*, uses the increment operator to increment the variable *count's* value by 1:

```
#include <iostream.h>

void main(void)
 {
   int count = 1000;

   cout << "count's starting value is " << count << endl;
   count++;
   cout << "count's ending value is " << count << endl;
 }
```

The *Inc_Op.CPP* program works exactly the same as *IncCount.CPP*, which used the assignment operator to increment the variable's value. When C++ encounters an increment operator, C++ first gets the variable's value, adds 1 to the value, and then stores the result back into the variable.

UNDERSTANDING PREFIX (BEFORE) AND POSTFIX (AFTER) INCREMENT OPERATORS

When you use the increment operator, you can place the operator before or after the variable, as shown here:

```
++variable;                              variable++;
```

In the first case, the operator appears in front of the variable, making the operator a *prefix increment operator*. In the second case, the operator appears after the variable, and is a *postfix increment operator*. As you program, you must understand that C++ treats these two operators differently. For example, consider the following assignment statement:

```
current_count = count++;
```

The assignment statement directs C++ to assign *count's* current value to the variable *current_count*. In addition, the postfix increment operator tells C++ to then increment *count's* current value. Using the postfix operator in this case makes the previous statement equivalent to the following two statements:

```
current_count = count;
count = count + 1;
```

Next, consider the following assignment statement that uses the prefix increment operator:

```
current_count = ++count;
```

In this case, the prefix operator tells C++ to *first* increment count's value and *then* assign the result to the variable *current_count*. Using the prefix increment operator makes the previous statement equivalent to the following two statements:

```
count = count + 1;
current_count = count;
```

It is important that you understand the prefix and postfix increment operators because you will see them in most C++ programs. The following program, *Pre_Post.CPP*, illustrates how to use the prefix and postfix increment operators:

```
#include <iostream.h>

void main(void)
 {
   int small_count = 0;
   int big_count = 1000;

   cout << "small_count is " << small_count << endl;
   cout << "small_count++ yields " << small_count++ << endl;
   cout << "small_count ending value " << small_count << endl;

   cout << "big_count is " << big_count << endl;
   cout << "++big_count yields " << ++big_count << endl;
   cout << "big_count ending value " << big_count << endl;
 }
```

When you compile and execute the *Pre_Post.CPP* program, your screen will display the following output:

```
C:\> Pre_Post    <ENTER>
small_count is 0
small_count++ yields 0
small_count ending value 1
big_count is 1000
++big_count yields 1001
big_count ending value 1001
```

The *Pre_Post.CPP* program uses the postfix increment operator with the variable *small_count*. As a result, the program displays the variable's current value (0) and then increments the value by 1. The program uses the prefix increment operator with the variable *big_count*. As a result, the program first increments the variable's value (1000 + 1) and then displays the result (1001). Take time now to edit this program to change the first postfix operator to a prefix operator and the second prefix operator to postfix. Compile and execute the program, and note how changing the operators changes the program's output.

C++ ALSO PROVIDES A DECREMENT OPERATOR

As you have just learned, the double plus sign (++) is the C++ increment operator. In a similar way, the double minus sign (--) is the C++ *decrement operator*, which decrements a variable's value by 1. As was the case with the increment operator, C++ supports a prefix and postfix decrement operator. The following program, *DecCount.CPP*, illustrates how to use the C++ decrement operator:

```cpp
#include <iostream.h>

void main(void)
 {
   int small_count = 0;
   int big_count = 1000;

   cout << "small_count is " << small_count << endl;
   cout << "small_count-- yields " << small_count-- << endl;
   cout << "small_count ending value " << small_count << endl;

   cout << "big_count is " << big_count << endl;
   cout << "--big_count yields " << --big_count << endl;
   cout << "big_count ending value " << big_count << endl;
 }
```

When you compile and execute the *DecCount.CPP* program, your screen will display the following output:

```
C:\> DecCount    <ENTER>
small_count is 0
small_count-- yields 0
small_count ending value is -1
big_count is 1000
--big_count yields 999
big_count ending value is 999
```

As you can see, the C++ prefix and postfix decrement operators work just as their increment operator counterparts, with the difference that they decrement the variable's value by 1.

OTHER C++ OPERATORS

This lesson has focused on the common C++ arithmetic operators and the increment and decrement operators. As you examine C++ programs, you may encounter one or more of the operators listed in Table 6.2.

Operator	Function
%	Modulo operator; returns the remainder of an integer division
~	One's complement operator; inverts a value's bits
&	Bitwise *AND* operator; *AND*s the ones bits between two values
|	Bitwise *OR* operator; *OR*s the ones bits between two values
^	Bitwise exclusive *OR*; exclusive *OR*s the bits between two values
<<	Bitwise left shift; shifts a value's bits left the number of positions specified
>>	Bitwise right shift; shifts a value's bits right the number of positions specified

Table 6.2 C++ operators you might encounter.

UNDERSTANDING OPERATOR PRECEDENCE

When you perform arithmetic operations in C++, you must be aware that C++ performs operations in a specific order, based on an *operator precedence*—meaning C++ considers some operators more important than others and C++ will perform the important operations first. For example, based on its operator precedence, C++ will perform a multiplication operation before it will perform an addition. To better understand operator precedence, consider the following expression:

```
result = 5 + 2 * 3;
```

Depending on the order that C++ performs the multiplication and addition operations, different results will occur, as shown here:

```
result = 5 + 2 * 3;        result = 5 + 2 * 3;
       = 7 * 3;                   = 5 + 6;
       = 21;                      = 11;
```

To avoid such mix-ups, C++ assigns a precedence to each operator that determines the order in which C++ performs operations. Because C++ performs operations in a consistent order, your programs will perform arithmetic calculations in a consistent manner.

Table 6.3 lists the C++ operator precedence. The operators that appear in the top section have the highest precedence. Within each section, operators have the same precedence. If you examine the table, you will find that C++ assigns a higher precedence to multiplication than to addition. You have not seen many of the operators that appear in the sections. For now, don't worry about these operators. By the time you finish this book, you will have used (and should understand) each of them.

Operator	Name	Example
::	Scope resolution	class_name::class_member_name
::	Global resolution	::variable_name
.	Member selector	object.member_name
–>	Member selector	pointer->member_name
[]	Subscript	pointer[element]
()	Function call	expression(parameters)
()	Value construction	type(parameters)
sizeof	Size of an object	sizeof expression
sizeof	Size of a type	sizeof(type)
++	Postfix increment	variable++
++	Prefix increment	++variable
--	Postfix decrement	variable--
--	Prefix decrement	--variable
&	Address operator	&variable
*	Dereference operator	*pointer
new	Allocate operator	new type
delete	Deallocate operator	delete pointer
delete[]	Deallocate array	delete pointer
~	Ones complement	~expression
!	NOT operator	! expression
+	Unary plus	+1
-	Unary minus	–1
()	Cast operator	(type) expression
.*	Member selector	object.*pointer
–>	Member selector	object–>*pointer
*	Multiply	expression * expression
/	Divide	expression / expression
%	Modulo	expression % expression
+	Addition	expression + expression
–	Subtraction	expression – expression

Table 6.3 C++ *Operator Precedence.*

CONTROLLING THE ORDER IN WHICH C++ PERFORMS OPERATIONS

As you have learned, C++ assigns a different precedence to operators, which controls the order in which C++ performs operations. Unfortunately, there may be times when the order in which C++ performs arithmetic operations does not match the order you want. For example, assume that your program must add two prices and then multiply the result by a tax rate, as shown here:

```
cost = price_a + price_b * 1.06;
```

Unfortunately, in this case, C++ will perform the multiplication first (price_b * 1.06) and then will add the value of *price_a*.

When your programs must perform arithmetic operations in a specific order, you can place expressions within parentheses. When C++ evaluates expressions, C++ always first performs the operations your program groups within parentheses. For example, consider the following expression:

```
result = (2 + 3) * (3 + 4);
```

C++ will evaluate this expression as shown here:

```
result = (2 + 3) * (3 + 4);
       = (5) * (3 + 4);
       =  5 * (7);
       = 5 * 7;
       = 35;
```

By grouping expressions within parentheses in this way, you can control the order in which C++ performs arithmetic operations. Given the previous example, your program can add the two prices within parentheses, as shown here:

```
cost = (price_a + price_b) * 1.06;
```

BE AWARE OF OVERFLOW WITH ARITHMETIC OPERATIONS

In Lesson 5, you learned that when you assign a value to a variable that falls outside the range of values the variable's type can store, an overflow error occurs. When you perform arithmetic operations, you must keep overflow errors in mind. For example, the following program, *MathOver.CPP*, multiplies the value 200 by 300 and assigns the result to a variable of type *int*. However, because the result of the multiplication (60,000) exceeds the largest value a variable of type *int* (32,767) can store, an overflow error occurs:

```
#include <iostream.h>

void main(void)
 {
   int result;

   result = 200 * 300;

   cout << "200 * 300 = " << result << endl;
 }
```

When you compile and execute the *MathOver.CPP* program, your screen will display the following output:

```
C:\> MathOver   <ENTER>
200 * 300 = -5536
```

WHAT YOU MUST KNOW

In this lesson, you examined the common C++ arithmetic and increment operators. As you learned, to ensure that your programs perform arithmetic calculations consistently, C++ assigns a precedence to each operator that controls the order in which C++ performs each operation. In Lesson 7, "Reading Keyboard Input," you will learn how to use an input stream named *cin* to perform keyboard input operations. Before you continue with Lesson 7, however, make sure that you have learned the following key concepts:

☑ C++ uses the operators +, –, *, and / for addition, subtraction, multiplication, and division.

☑ C++ provides prefix (before) and postfix (after) increment operators that add 1 to a variable's value.

☑ C++ provides prefix (before) and postfix (after) decrement operators that subtract 1 from a variable's value.

☑ The C++ prefix (before) operators direct C++ to first increment (or decrement) the variable's value and then to use the value.

☑ The C++ postfix (after) operators direct C++ to first use the variable's value and then to increment (or decrement) the value.

☑ To ensure that C++ evaluates expressions consistently, C++ assigns a precedence to each operator that controls the order in which operations occur.

☑ When you must control the order in which arithmetic operations are performed, place your expressions in parentheses. C++ always evaluates expressions in parentheses first.

LESSON 7

READING KEYBOARD INPUT

Throughout this book, your programs have made extensive use of the *cout* output stream to display output to your screen. In this lesson, you will learn that C++ provides an input stream named *cin* from which your programs can read information a user types at the keyboard. When you use *cin* to read keyboard input, you specify one or more variables to which you want *cin* to assign the input values. By the time you finish this lesson, you will understand the following key concepts:

- C++ programs can use the *cin* input stream to read and assign letters and numbers a user types at the keyboard to specific variables.

- After you use *cin* to read and assign keyboard input to a variable, you can use the variable's contents just as if your program had used the assignment operator to store the value within the variable.

- When your program uses *cin* to perform keyboard input, your program must consider overflow errors and errors that occur when the user enters a value of the wrong type (a value that does not match the kind of value the variable can store).

As you have learned, when your programs use the *cout* output stream, your programs place data into the stream using the *insertion* operator (<<). In a similar way, when your programs use *cin* to read keyboard input, your programs will use the *extraction* operator (>>).

GETTING STARTED WITH CIN

Just as the *cout* output stream lets your programs write output to the screen display, the *cin* input stream lets your program read input from the keyboard. When your programs use *cin* to read keyboard input, your programs must specify a variable into which *cin* will place the data. The following program, *FirstCin.CPP*, uses *cin* to read a number a person types at a keyboard. The program assigns the number the user types at the keyboard to a variable named *number* and then displays the variable's value using the *cout* output stream, as shown here:

```
#include <iostream.h>

void main(void)
 {
   int number;    // The number read from the keyboard

   cout << "Type your favorite number and press Enter: ";
   cin >> number;
   cout << "Your favorite number is " << number << endl;
 }
```

When you compile and execute the *FirstCin.CPP* program, your screen will display a message prompting you to type in your favorite number. When you type the number and press ENTER, the program will assign your input to the variable *number*. Using *cout*, the program will then display a message containing your favorite number.

The following program, *TwoNbrs.CPP*, prompts you to type in two numbers. The program assigns the numbers to the variables *first* and *second*. The program then displays the numbers using *cout*, as shown here:

```
#include <iostream.h>

void main(void)
 {
   int first, second;  // Numbers typed at the keyboard

   cout << "Type two numbers and press Enter: ";
   cin >> first >> second;
   cout << "The numbers typed were " << first << " and " << second << endl;
 }
```

Note the use of the two extraction operators with *cin*:

```
cin >> first >> second;
```

In this case, *cin* will assign the first value typed to the variable *first* and the second to the variable *second*. If your program requires a third value, you can use a third extraction operator, as shown here:

```
cin >> first >> second >> third;
```

When you use *cin* to read characters from the keyboard, *cin* will use the first *whitespace character* (a space, tab, or carriage return) to determine where one value ends and the second begins. Experiment with the *TwoNbrs* program, separating the numbers with a space, tab, or carriage return. For example, the first time you run the program, press the TAB key between the numbers. The second time you run the program, press ENTER after each number.

Reading Keyboard Input Using cin

When your programs must read input from the keyboard, your programs can use the *cin* input stream. When your programs use *cin*, you must use the extraction operator (>>) to specify the variable into which you want *cin* to place the data, as shown here:

```
cin >> some_variable;
```

The extraction operator is so named because the operator extracts (removes) data from an input stream and assigns the data to the variable that appears to the right of the operator.

BE AWARE OF OVERFLOW ERRORS

When your programs perform their input using *cin*, you must be aware of possible errors that can occur when the user types an invalid number. For example, run the program *FirstCin*, which you just created. When the program prompts you to type your favorite number, type the number 1000000 and press ENTER. The program will not display the number 1000000 as the value typed. Instead, because the value 1000000 exceeds the largest value a value of type *int* can store, an overflow error will occur.

If you examine the program *FirstCin.CPP*, you will note that *cin* assigns the number typed to a variable of type *int*. As you learned in Lesson 5, "Programs Store Information in Variables," variables of type *int* can only hold values in the range −32,768 through 32,767. Because a variable of type *int* cannot hold the value 1000000, the program generated an overflow error. Run the program a few more times, typing in negative and positive numbers. Note the errors that occur when you exceed the acceptable range of values for the variable into which *cin* places the input.

BE AWARE OF TYPE MISMATCH ERRORS

As discussed, the program *FirstCin.CPP* expects the user to type a value in the range -32,768 through 32,767. If rather than typing a value outside of this range, the user types characters or other symbols, an *type-mismatch error* will occur. In other words, the program expects a value of one type (*int*) and the user enters values of a second type (*char*). As an example, run the program a second time. When the program prompts you for a number, type in the letters *ABC*. As before, because program expects an integer value, not letters, an error occurs.

Perform similar experiments with the program *TwoNbrs*, possibly typing in nonsense values or even floating-point numbers. As you will find, as you type in errant values, the program's output will also contain errors. In later lessons you will learn how to perform input operations in ways that minimize the possibility of such errors. For now, however, simply be aware that such errors can occur.

READING CHARACTER DATA

The previous programs both used *cin* to read integer numbers into variables of type *int*. The following program, *Cin_Char.CPP*, uses the *cin* input stream to read a keyboard character. As you can see, the program reads the character into a variable of type *char*:

```
#include <iostream.h>

void main(void)
 {
   char letter;

   cout << "Type any character and press Enter: ";
   cin >> letter;
   cout << "The letter typed was " << letter << endl;
 }
```

Compile and experiment with the *Cin_Char.CPP* program, possibly typing more than one character and watching the program's response. As you will find, the program works with only one character at a time.

READING VALUES FROM THE KEYBOARD

In Section 2 of this book you will learn how to store words, or even a line of text, within a variable. At that time, you will learn how to use the *cin* input stream to read words and complete lines. For now, however, you may want to create your own simple programs that read values of type *float* or *long*. For example, the following program, *Cin_Long.CPP*, uses *cin* to read a *long* value:

```
#include <iostream.h>

void main(void)
 {
   long value;
```

```
    cout << "Type a large number and press Enter: ";
    cin >> value;
    cout << "The number you typed was " << value << endl;
}
```

As before, experiment with the *Cin_Long.CPP* program by typing in large (and negative) numbers.

I/O Redirection and the cin Input Stream

As you learned in Lesson 4, "Writing Messages to Your Screen," when your programs use the *cout* output stream, a user can redirect your program's output from the screen display to a file or printer. As discussed, the *cout* output stream corresponds to the operating system's standard output. In a similar way, the *cin* input stream corresponds to the operating system's standard input. As a result, when your program uses *cin* to perform input operations, a user can redirect the program's input from the keyboard to a file. In later lessons you will learn how to write programs that read and process redirected input.

What You Must Know

In this lesson you learned how to use the *cin* input stream to perform keyboard input. As you learned, when your programs use *cin* to read keyboard input, you must specify the variables to which *cin* assigns the values the user types. In Lesson 8, "Teaching Your Program to Make Decisions," you will learn how to use the C++ *if* statement to let your programs make their own decisions. Before you continue with Lesson 8, however, make sure that you have learned the following key concepts:

- ☑ C++ provides the *cin* input stream your programs can use to read keyboard input.

- ☑ When your programs use *cin* to read input, your programs must specify one or more variables into which *cin* will place the data.

- ☑ To assign input to a variable, you must use *cin* with the extraction operator (>>).

- ☑ When your programs use *cin* to read multiple values, *cin* uses whitespace characters (a space, tab, or carriage return) to determine where one value ends and a second begins.

- ☑ If a user does not type valid data, overflow or type mismatch errors can occur, and the values *cin* assigns to your program variables will be in error.

LESSON 8

TEACHING YOUR PROGRAM TO MAKE DECISIONS

As you have learned, a program is a list of instructions the computer performs to accomplish a specific task. All the simple C++ programs you have created so far have started with the first statement in the program and have executed each statement, in order, to the end of the program. As your programs become more complex, there will be times when you will want the programs to execute one set of statements if one condition is true and, possibly, another set if the condition is false. In other words, you will want your programs to make decisions and respond accordingly. This lesson examines the C++ *if* statement that your programs will use to make such decisions. By the time you finish this lesson, you will understand the following key concepts:

- C++ programs use *relational operators* to determine if two values are equal or if one value is larger or smaller than the other.

- C++ programs use the *if* statement to make decisions.

- C++ statements can be *simple* (one operation) or *compound* (multiple operations that your program groups within right and left braces {}).

- C++ programs use the *if-else* statement to perform one set of statements when a condition is true and a second set of statements if the condition is false.

- By combining several *if-else* statements, your programs can test for and respond to different conditions.

- Using the C++ *AND* (&&) and *OR* (||) operators, your programs can test for multiple conditions, such as *Does the user have a dog AND is the dog a Dalmatian?*

Programs that make decisions perform *conditional processing*. In other words, based on the outcome of one or more conditions, the program will execute specific statements. Experiment with the programs this lesson presents. Your collection of C++ tools is now becoming large enough for you to create useful programs.

COMPARING TWO VALUES

To make decisions, your programs must first perform some type of test. For example, one program might test if a student's test score is equal to 100 and a second program might test if the cost of an item is more than $50.00. To perform such tests, your programs will use the C++ relational operators listed in Table 8.1. *Relational operators* let your programs test how one value "relates" to another. In other words, using relational operators, your programs can test whether one value is equal to, greater than, or less than a second value. When your programs use relational operators to compare two values, the result of the comparison is either true or false—meaning, the two values are either equal (true) or they are not equal (false). Each of the *if* statements the programs this book presents will use the relational operators listed in Table 8.1.

Operator	Test	Example
==	If two values are equal	(score == 100)
!=	If two values are not equal	(old != new)

Table 8.1 The C++ relational operators. (continued on following page)

Operator	Test	Example
>	If the first value is greater than the second	(cost > 50.00)
<	If the first value is less than the second	(salary < 20000.00)
>=	If the first value is greater than or equal to the second	(stock_price >= 30.0)
<=	If the first value is less than or equal to the second	(age <= 21)

Table 8.1 *The C++ relational operators. (continued from previous page)*

GETTING STARTED WITH THE IF STATEMENT

The C++ *if* statement lets your programs perform a test and then execute statements based on the result of the test. The format of the *if* statement is as follows:

```
if (condition_is_true)
   statement;
```

The *if* statement normally performs a test using a C++ relational operator. If the test result is true, the program executes the statement that follows the *if*. On the other hand, if the test result is false, the program ignores (skips) the statement that follows. The following program, *First_If.CPP*, uses the *if* statement to compare the value stored in the variable *test_score* to value 90. If the test score is greater than or equal to 90, the program will display a message that tells the user he or she got an A. Otherwise, if the value is less than 90, the program will simply end:

```
#include <iostream.h>

void main(void)
  {
    int test_score = 95;

    if (test_score >= 90)
      cout << "Congratulations, you got an A!" << endl;
  }
```

As you can see, the program uses the C++ greater-than-or-equal-to relational operator (>=) to perform the test. If the value comparison results in true, the program will execute the statement that follows—in this case, displaying the message using *cout*. If the comparison results in false, the program will not display the message. Experiment with this program, changing the test score to a value that is less than 90, and note the processing the *if* statement performs.

UNDERSTANDING SIMPLE AND COMPOUND STATEMENTS

When your programs use the *if* statement for conditional processing, there will be times when your programs must only perform one statement if the condition is true. At other times, your program must perform several statements when a condition is true. When your program performs only one statement following an *if*, the statement is a *simple statement*, as shown here:

```
if (test_score >= 90)
   cout << "Congratulations, you got an A!" << endl;        ———— Simple statement
```

For your program to perform several statements when a condition evaluates as true, you must group the statements within left and right braces {}. The statements that appear within the braces make up a *compound statement*, as shown here:

```
if (test_score >= 90)
   {
     cout << "Congratulations, you got an A!" << endl;        ———— Compound
     cout << "Your test score was " << test_score << endl;         statement
   }
```

It is not important that you remember the terms simple and compound statements, but rather, that you know that you must group related statements within the left and right braces. The following program, *Compound.CPP*, changes the previous program to display two messages if the test score is greater than or equal to 90:

```
#include <iostream.h>

void main(void)
 {
   int test_score = 95;

   if (test_score >= 90)
     {
       cout << "Congratulations, you got an A!" << endl;
       cout << "Your test score was " << test_score << endl;
     }
 }
```

Using Simple and Compound Statements

When your programs perform conditional processing, there will be times when they must perform only one statement (a simple statement) when a condition is true. At other times, however, your programs must perform multiple statements (a compound statement). When your programs must perform two or more related statements based on a condition, you must group the statements within left and right braces, as shown here:

```
if (age >= 21)
  {
    cout << "Make sure you remember to vote!" << endl;
    cout << "Oh yeah, this Bud's for you!" << endl;
  }
```

PROVIDING ALTERNATIVE STATEMENTS FOR FALSE CONDITIONS

The previous two programs used an *if* statement to determine whether a test score was greater than or equal to 90. If the condition was true, the programs displayed messages to the screen. If the condition was false, meaning the test score was less than 90, the program did not display a message, it simply ended. In most cases, your programs must specify one set of statements that executes when the condition is true and a second set that executes if the condition is false. To provide the statements that execute when the condition is false, your programs must use the *else* statement. The format of the *else* statement is as follows:

```
if (condition_is_true)
    statement;
else
    statement;
```

The following program, *If_Else.CPP*, uses the *if* statement to test whether the test score is greater than or equal to 90. If the condition is true, the program will display a message of congratulations. If the condition is false, the program will display a message telling the student to work harder, as shown here:

```
#include <iostream.h>

void main(void)
 {
   int test_score = 95;

   if (test_score >= 90)
      cout << "Congratulations, you got an A!" << endl;
   else
      cout << "You must work harder next time!" << endl;
 }
```

COMPOUND STATEMENTS APPLY TO ELSE, TOO

As you have learned, a compound statement groups related statements within left and right braces. When your program uses an *else* statement to specify statements the program will perform when a condition is false, you can use a compound statement to group multiple statements. The following program, *Cmp_Else.CPP*, uses a compound statement for both the *if* and the *else*:

```
#include <iostream.h>

void main(void)
 {
   int test_score = 65;

   if (test_score >= 90)
      {
        cout << "Congratulations, you got an A!" << endl;
        cout << "Your test score was " << test_score << endl;
      }
   else
      {
        cout << "You should have worked harder!" << endl;
        cout << "You missed " << 100 - test_score <<  " points " << endl;
      }
 }
```

55

As before, take time to experiment with this program, changing the *test_score* variable to values less than and greater than 90. The following program, *GetScore.CPP*, uses the *cin* input stream to get the test score from the user. The program then compares the test score to 90, displaying the corresponding messages:

```cpp
#include <iostream.h>

void main(void)
 {
   int test_score;

   cout << "Type in the test score and press Enter: ";
   cin >> test_score;

   if (test_score >= 90)
     {
       cout << "Congratulations, you got an A!" << endl;
       cout << "Your test score was " << test_score << endl;
     }
   else
     {
       cout << "You should have worked harder!" << endl;
       cout << "You missed " << 100 - test_score << " points " << endl;
     }
 }
```

As you can see, the program uses the *cout* output stream to prompt the user to enter a test score. Next, the program uses the *cin* input stream to assign the user's response to the variable *test_score*. Compile and execute the *GetScore.CPP* program. As you will find, when you combine input operations with conditional processing, your programs will become very powerful.

USE INDENTATION TO IMPROVE YOUR PROGRAM'S READABILITY

If you examine the programs this chapter presents, you will find that the programs indent the statements that follow an *if*, *else*, or left brace. By indenting your statements one or two spaces in this way, you make it easy for someone who is reading your program to determine how statements relate, as shown here:

```cpp
   if (test_score >= 90)
     {
       cout << "Congratulations, you got an A!" << endl;
       cout << "Your test score was " << test_score << endl;
     }
   else
     {
       cout << "You should have worked harder!" << endl;
       cout << "You missed " << 100 - test_score << " points " << endl;
     }
```

In this case, by simply noting the indentation, another programmer who is reading your code can quickly identify which statements relate to the *if* and which statements relate to the *else*. As you create your programs, use similar indentation to make your programs more readable. C++ does not care about the indentation, but programmers who are reading and trying to understand your code will.

Understanding if-else Processing

As your programs become more complex, your programs will test different conditions and perform one set of statements when the condition is true and a second set when the condition is false. To perform such conditional processing, your programs will use *if-else* statements, as shown here:

```
if (condition_is_true)
    statement;
else
    statement;
```

If your programs must perform more than one statement when the condition is true or false, you must group the related statements within left and right braces {}, as shown here:

```
if (condition_is_true)
  {
    first_true_statement;
    second_true_statement;
  }
else
  {
    first_false_statement;
    second_false_statement;
  }
```

TESTING TWO OR MORE CONDITIONS

As you have learned, the *if* statement lets your program test specific conditions. As your programs become more complex, there will be times when you will test more than one condition. For example, your program might test whether a test score is greater than or equal to 90 and whether a student's grade is currently an A. Likewise, you might test whether a user owns a dog and whether that dog is a Dalmatian. To perform such operations, you will use the C++ logical *AND* operator (&&). In addition, if you want to test whether a user owns a dog or a cat, you would use the logical OR operator (||). When your programs use the logical *AND* or logical *OR* operator to test more than one condition, you will place each condition within parentheses, as shown here:

```
if ((user_owns_a_dog) && (dog == dalmatian))
```
— Entire condition

As you can see, the program groups each condition within its own parentheses, which the program then groups within an outer set of parentheses.

```
if ((user_owns_a_dog) && (dog == dalmatian))
```

When your programs use the logical *AND* operator (&&), all the conditions within the statement must be true for the entire condition to evaluate as true. If any condition is false, the entire condition becomes false. For example, if the user does not own a dog, the previous condition is false. Likewise, if the user's dog is not a Dalmatian, the condition is false. In order for the condition to be true, the user must own a dog and that dog must be a Dalmatian.

The following statement uses the logical *OR* operator (||) to determine if a user owns a dog or cat:

```
if ((user_owns_a_dog) || (user_owns_a_cat))
```

For a condition that uses the logical *OR* operator to evaluate as true, only one condition must be true. For example, if the user owns a dog, the previous condition is true. If the user owns a cat, the condition is true. Also, if the user owns both a dog and a cat, the condition is true. The only time the condition would be false is if the user does not own a dog or a cat.

C++ REPRESENTS TRUE AS ANY NONZERO VALUE AND FALSE AS ZERO

Many C++ programs take advantage of the fact that C++ represents true using any nonzero value, and false as zero. For example, assume your program uses a variable named *user_owns_a_dog* to determine whether or not the user owns a dog. If the user does not own a dog, you can assign the value 0 (false) to the variable as shown here:

```
user_owns_a_dog = 0;
```

If the user owns a dog, you can assign any nonzero value to the variable, such as the value 1:

```
user_owns_a_dog = 1;
```

Your programs can then test the variable using an *if* statement, as shown here:

```
if (user_owns_a_dog)
```

If the variable contains a nonzero value, the condition will evaluate as true; otherwise, if the variable contains the value 0, the condition is false. By taking advantage of how C++ represents true and false, the previous statement is identical to the following:

```
if (user_owns_a_dog == 1)
```

The following program, *Dog_Cat.CPP*, uses the variables *user_owns_a_dog* and *user_owns_a_cat* within *if* statements to determine the types of animals the user owns:

```
#include <iostream.h>

void main(void)
  {
    int user_owns_a_dog = 1;
    int user_owns_a_cat = 0;

    if (user_owns_a_dog)
      cout << "Dogs are great" << endl;

    if (user_owns_a_cat)
      cout << "Cats are great" << endl;
```

```
    if ((user_owns_a_dog) && (user_owns_a_cat))
      cout << "Dogs and cats can get along" << endl;

    if ((user_owns_a_dog) || (user_owns_a_cat))
      cout << "Pets are great!" << endl;
  }
```

Experiment with this program by assigning the value 1 to both variables, 0 to both variables, and then 1 and 0 to different variables. Using the logical *AND* and *OR* operators, testing two or more conditions within your programs is very easy.

USING THE C++ *NOT* OPERATOR

As you have learned, when your programs test for specific conditions, there are times when you want the program to perform specific statements when a condition is true. In a similar way, there may be times when you want your programs to perform a set of statements when a condition is not true. The C++ *NOT* operator, the exclamation point (!), lets your programs test if a condition is *not* true. For example, the following statement tests whether the user does not own a dog:

```
  if (! user_owns_a_dog)
    cout << "You should buy a dog" << endl;
```

The *NOT* operator converts a false condition to true and a true condition to false. For example, assume that the user does not own a dog. The variable *user_owns_a_dog* would contain the value 0. When C++ performs the condition using the *NOT* operator, C++ uses the variable's current value (0), and applies the *NOT* operator. The *NOT* operator makes the 0 value 1 (true). The entire condition then evaluates as true, and the program performs corresponding statements.

The following program, *Use_Not.CPP*, illustrates how to use the *NOT* operator:

```
  #include <iostream.h>

  void main(void)
   {
     int user_owns_a_dog = 0;
     int user_owns_a_cat = 1;

     if (! user_owns_a_dog)
       cout << "You should buy a dog" << endl;

     if (! user_owns_a_cat)
       cout << "You should buy a cat" << endl;
   }
```

As before, experiment with the values you assign to the variables *user_owns_a_dog* and *user_owns_a_cat* and watch the processing the program performs. As your programs become more complex, they will use the *NOT* operator on a regular basis. For example, your program may continue to repeat its processing as long as it has *not* encountered the end of a file.

Using C++ Logical Operators

As you specify conditions within your programs, there will be times when the conditions have multiple parts. For example, your program might test if an employee is paid hourly and has worked over 40 hours this week. When your conditions require two parts to be true for the condition to be true, you will use the C++ *AND* operator (&&). To use the *AND* operator, group each condition within its own parentheses and both conditions within their own parentheses, as shown here:

```
if ((employee_pay == hourly) && (employee_hours > 40))
  statement;
```

When your condition requires only one of two parts to be true for the condition to be true, your programs should use the C++ *OR* operator (||). For example, the following condition tests whether the user owns either a car or motorcycle:

```
if ((vehicle == car) || (vehicle == motorcycle))
  statement;
```

As before, the program groups each condition within parentheses. In some cases, you may want your programs to perform a statement when a condition is not true. In such cases, you should use the C++ *NOT* operator (!). The *NOT* operator converts a true condition to false and a false condition to true. The C++ *AND*, *OR*, and *NOT* operators are *logical operators*.

HANDLING DIFFERENT CONDITIONS

This lesson's programs have used *if* and *else* to specify one set of statements the program is to perform when a condition is true, and another set of statements the program will perform if the condition is false. There may be times, however, when your programs must test several different related conditions. For example, assume that your program must determine a student's test grade. To do so, your program must test for the scores greater than or equal to 90, 80, 70, 60, and so on. The following program, *ShowGrad.CPP*, uses a series of *if-else* statements to determine a grade:

```
#include <iostream.h>

void main(void)
 {
   int test_score;

   cout << "Type in your test score and press Enter: ";
   cin >> test_score;

   if (test_score >= 90)
     cout << "You got an A!" << endl;
   else if (test_score >= 80)
     cout << "You got a B!" << endl;
   else if (test_score >= 70)
     cout << "You got a C" << endl;
```

```
    else if (test_score >= 60)
      cout << "Your grade was a D" << endl;
    else
      cout << "You failed the test" << endl;
}
```

When the program performs the first *if* statement, it tests whether the test score is greater than or equal to 90. If so, the program will display a message telling the user he or she received an A. If the test score is not greater than or equal to 90, the program performs the following *else if* to test whether the score is greater than or equal to 80. The program repeats this processing until it determines the correct grade. As before, experiment with this program, typing in different test scores. Also, note that in this case, the final *else* statement is not an *if-else*. If the student did not receive an A, B, C, or D, the student must have failed the test. Therefore, there is no reason for the program to perform an additional test.

USING THE SWITCH STATEMENT

As you just learned, by combining a series of *if-else* statements, your programs can test multiple conditions. In the previous program, you used *if-else* statements to determine if a test score fell within a range of values. For cases in which your programs must test for specific values, your programs can use the C++ *switch* statement.

When you use the *switch* statement, you must specify a condition and then one or more cases the program will try to match to the condition. For example, the following program, *Switch.CPP*, uses a *switch* statement to display a message based on a student's current grade:

```
#include <iostream.h>

void main(void)
  {
    char grade = 'B';

    switch (grade) {
      case 'A': cout << "Congratulations on your A" << endl;
              break;
      case 'B': cout << "Not bad, a B is ok" << endl;
              break;
      case 'C': cout << "C's are only average" << endl;
              break;
      case 'D': cout << "D's are terrible" << endl;
              break;
      default: cout << "No excuses! Study harder!" << endl;
              break;
    }
  }
```

The *switch* statement consists of two parts. The first part of the *switch* statement specifies the condition that appears after the keyword *switch*. The second part specifies the possible matching cases. When the program encounters a *switch* statement, the program first examines the condition and then tries to find a matching value within the possible cases. When program finds a match, the program executes the corresponding statements. In the case of the previous program, the case for a letter grade of B matches the condition. Therefore, the program will display a message telling the user that a B is not bad. Take time to experiment with this program, changing the letter grade and watching the corresponding processing. The *default* case provides a "catch all" case that will match any condition.

Note the use of the *break* statement with each case in the previous program. As it turns out, when C++ encounters a case that matches the condition in a *switch* statement, C++ considers all the cases that follow a match as well. The *break* statement tells C++ to end the current *switch* statement and to continue the program's execution at the first statement that follows the *switch* statement. If you remove the *break* statements from the previous program, the program will display a message not only for the matching case, but for each of the cases that follow (because C++ considers all cases as true after one case is true).

What You Must Know

In this lesson you learned how to use the C++ *if* statement to perform conditional processing, which lets your programs make their own decisions. As you have learned, your programs can use the *if* statement to perform one set of statements when a condition is true and the *else* statement to specify statements the program executes if the condition is false. In Lesson 9, "Repeating One or More Statements," you will learn how to use C++ iterative statements to repeat statements a specific number of times or until a specific condition occurs. For example, you might repeat the same statements 100 times to add up 100 student test scores. Before you continue with Lesson 9, however, make sure that you have learned the following key concepts:

- ☑ C++ relational operators let your programs test whether two values are equal or not equal, or if one value is greater than or less than another.

- ☑ The C++ *if* statement lets your program test a condition and perform one or more statements if the condition is true.

- ☑ The C++ *else* statement lets your programs specify one or more statements that execute when a condition an *if* statement tests is false.

- ☑ C++ represents true using any nonzero value, and false as 0.

- ☑ The C++ logical *AND* (&&) and *OR* (||) operators let your programs test for more than one condition.

- ☑ The C++ logical *NOT* (!) operator lets your programs test for conditions that are not true.

- ☑ If your programs must execute more than one statement for an *if* or *else*, you must place the statements within left and right braces {}.

- ☑ Indent your program statements to help programmers who read your code determine related statements quickly.

- ☑ When your programs must test if a condition matches specific values, your programs can use the *switch* statement.

- ☑ When your program encounters a matching case within a *switch* statement, C++ considers all the cases that follow as a match as well. Using the *break* statement, you can instruct C++ to end the *switch* statement and to continue the program's processing at the first statement that follows the *switch* statement.

LESSON 9

REPEATING ONE OR MORE STATEMENTS

In Lesson 8, "Teaching Your Program to Make Decisions," you learned how to use the C++ *if* statement to make decisions within your programs. Closely related to such decision making within your program is the ability to repeat one or more instructions a specific number of times or until a known condition occurs, such as the program reaching the end of a data file. In this lesson you will use the C++ looping (iterative) constructs to repeat one or more statements. Depending on your program's processing, you may use the *for, while,* or *do while* loops to repeat program statements. By the time you finish this lesson, you will understand the following key concepts:

- The C++ *for* statement lets your programs repeat statements a specific number of times.

- The C++ *while* statement lets your programs repeat statements as long as a known condition is true.

- The C++ *do while* statement lets your programs perform statements at least one time and then possibly repeat the statements based on a specified condition.

The ability to repeat statements is a very powerful programming feature. Experiment with the programs this lesson presents. After you have finished, you will have considerable C++ programming skills.

REPEATING STATEMENTS A SPECIFIC NUMBER OF TIMES

As you program, one of the most common operations your programs will perform is to repeat one or more statements a specific number of times. For example, one program might repeat the same statements to print five copies of a file. Likewise, a second program might repeat a set of statements 30 times to determine, for example, whether your 30 stocks are gaining or losing value. The C++ *for* statement makes it very easy for your programs to repeat one or more statements a specific number of times.

When your program uses a *for* statement (often called a *for loop*), your program must specify a variable, called a *control variable*, that keeps track of the number of times the loop executes. For example, the following *for* loop uses the variable *count* to keep track of the number of times the program has executed the loop. In the following example, the loop will execute ten times:

```
for (count = 1; count <= 10; count++)
    statement;
```

The *for* statement consists of four parts: an initialization, a test condition, the statements that are to repeat, and an increment. To begin, the statement *count = 1;* assigns the control variable's starting value. The *for* loop performs this initialization one time, when the loop first starts. Next, the loop tests the condition *count <= 10*. If the condition is true, the *for* loop will execute the statement that follows. If the condition is false, the loop will end, and the program will continue its execution with the first statement that follows the loop. If the condition is true and the *for* loop executes the statement, the loop will then increment the variable *count* using the statement *count++*. Finally, the program tests the condition *count <= 10*. If the condition is still true, the program will execute the statements, and the process of incrementing and then testing the variable *count* will repeat, as shown here:

```
for (count = 1; count <= 10; count++)
      └ Initialization  └ Test        └ Increment
```

The following program, *FirstFor.CPP*, uses the *for* loop to display the values 1 through 100 on your screen display:

```
#include <iostream.h>

void main(void)
 {
   int count;

   for (count = 1; count <= 100; count++)
     cout << count << ' ';
 }
```

As you can see, the *for* loop initializes the variable *count* to the value 1. The loop then tests if *count's* value is less than or equal to 100. If so, the *for* loop will execute the corresponding statement and then increment *count*, repeating the test. Experiment with the program, changing the value 100 to 10, 20, and even 5000.

The following program, *AskCount.CPP*, displays a message asking the user to type in the number at which the loop should end. The program then displays the numbers from 1 through the value the user has specified:

```
#include <iostream.h>

void main(void)
 {
   int count;
   int ending_value;

   cout << "Type in the ending value and press Enter: ";
   cin >> ending_value;

   for (count = 0; count <= ending_value; count++)
     cout << count << ' ';
 }
```

Experiment with the *AskCount.CPP* program, typing in values such as 10, 1, and even 0. If you type in the value 0 or −1, the *for* loop will never execute because the condition *count <= ending_value* will immediately fail. Remember, if you type in a value outside the range of values a variable of type *int* can store, an overflow error will occur. As an example, execute the program and type in the value 50000. Because the value exceeds the largest value a variable of type *int* can store, the overflow results in a negative value, which prevents the loop's execution.

C++ FOR LOOPS SUPPORT COMPOUND STATEMENTS

In Lesson 8, you learned that when your programs perform more than one statement within an *if* or an *else* statement, you must group the statements within left and right braces. The same is true for multiple statements and the *for* loop. The following program, *Add1_100.CPP*, loops through the numbers 1 through 100, displaying each number and then adding the number to a grand total:

```
#include <iostream.h>

void main(void)
 {
   int count;
   int total = 0;
```

```
   for (count = 1; count <= 100; count++)
     {
       cout << "Adding " << count << " to " << total;
       total = total + count;
       cout << " yields " << total << endl;
     }
 }
```

By grouping the statements within braces, the for loop can execute multiple statements with each loop (called an *iteration* of the loop).

CHANGING THE FOR LOOP'S INCREMENT

So far, each of the *for* loops in this lesson's programs has incremented the loop's control variable by 1 with each iteration of the loop. However, the *for* loop does not limit your programs to incrementing the variable by 1. The following program, *By_Fives.CPP*, displays every fifth number from 0 through 100:

```
#include <iostream.h>

void main(void)
  {
    int count;

    for (count = 0; count <= 100; count += 5)
      cout << count << ' ';
  }
```

When you compile and execute the *By_Fives.CPP* program, your screen will display the numbers 0, 5, 10, and so on, through 100. Note the statement the *for* loop uses to increment the variable *count*:

```
count += 5;
```

When you want to add a value to a variable's current value and then assign the result to the same variable, C++ lets your programs do so in one of two ways. First, assuming your program must add the value 5 to variable *count*, your program can do so as shown here:

```
count = count + 5;
```

Second, C++ lets you use the shorthand notation shown here to add the value 5 to the variable *count*:

```
count += 5;
```

Because it is easier to write, programmers commonly use this shorthand notation within loops. When you use a *for* loop, C++ does not limit your loops to counting up. The following program, *Cnt_Down.CPP*, uses a *for* loop to count down and display the numbers from 100 to 1:

```
#include <iostream.h>

void main(void)
  {
    int count;
```

```
   for (count = 100; count >= 1; count--)
     cout << count << ' ';
}
```

As you can see, the *for* loop initializes the variable *count* to 100. With each iteration, the loop decrements the variable's value by 1. When the variable *count* contains the value 0, the loop ends.

Be Aware of Infinite Loops

 As you have learned, a *for* loop provides your programs with a way to repeat related statements a specific number of times. Using a control variable, the *for* loop essentially counts the number of iterations it has performed. When the loop reaches its ending condition, your program stops repeating the statements and continues its execution at the first statement that follows the *for* loop.

Unfortunately, due to errors within programs, there are times when a loop never reaches its ending condition and, therefore, loops forever (or until you end the program). Such unending loops are called *infinite loops*. In other words, they are loops that have no way of ending. The following *for* statement, for example, creates an infinite loop:

```
   for (count = 0; count < 100; wrong_variable++)
     // Statements
```

As you can see, the *for* loop uses the variable *count* as its control variable. Within the loop's increment section, however, the program increments the wrong variable. As a result, the loop never increments the variable *count* and *count* will never have a value greater-than-or-equal-to 100. Thus, the loop becomes a never-ending infinite loop.

It is important to note that *for* loops are not restricted to using values of type *int* as their loop control variable. The following program, *LoopVar.CPP*, for example, uses a variable of type *char* to display the letters of the alphabet within one loop and a variable of type *float* to display floating-point numbers within a second loop:

```
#include <iostream.h>

void main(void)
  {
    char letter;
    float value;

    for (letter = 'A'; letter <= 'Z'; letter++)
      cout << letter;

    cout << endl;

    for (value = 0.0; value <= 1.0; value += 0.1)
      cout << value << ' ';

    cout << endl;
  }
```

When you compile and execute the *LoopVar.CPP* program, your screen will display the following output:

```
ABCDEFGHIJKLMNOPQRSTUVWXYZ
0 0.1 0.2 0.3 0.4 0.5 0.6 0.7 0.8 0.9
```

Looping a Specific Number of Times

As you program, one of the most common operations your programs will perform is to repeat one or more statements a specific number of times. The C++ *for* statement lets your programs do just that. The *for* statement uses a control variable that keeps track of the number of times the loop has executed. The general format of the *for* statement is as follows:

```
for (initialization; test; increment)
    statement;
```

When the *for* loop begins, it assigns (initializes) the starting value to the loop's control variable. Next, the program tests the loop's condition. If the condition is true, the program will execute the loop's statements. Finally, the program increments the loop's control variable and repeats the condition test. If the condition is true, the process will repeat. If the condition is false, the *for* loop will end and the program will continue its execution at the first statement that follows the *for* statement.

LOOPING WITH A WHILE LOOP

As you just learned, the C++ *for* loop lets your programs repeat one or more statements a specific number of times. In some cases, however, your programs must repeat statements as long as a specific condition is true. For example, in later lessons you will learn how to read a file's contents using a C++ program. Such programs may repeatedly loop until they encounter the end of the file. For situations in which your programs must loop as long as a specific condition is true, but not necessarily a specific number of times, your programs can use the C++ *while* statement. The general format of the *while* statement is shown here:

```
while (condition_is_true)
    statement;
```

When your program encounters a *while* statement, your program first tests the specified condition. If the condition is true, the program will execute the *while* loop's statements. After the last statement in the loop executes, the *while* loop again tests the condition. If the condition is still true, the loop's statements will repeat, and this process will continue. When the condition finally becomes false, the loop will end, and your program will continue its execution at the first statement that follows the loop.

The following program, *Get_YN.CPP*, prompts you to type Y for yes or N for no. The program then uses a *while* loop to read characters from the keyboard until the user types a Y or N. If the user types a value other than Y or N, the program will beep the computer's built-in speaker by writing the bell character '\a' to the *cout* output stream:

```
#include <iostream.h>

void main(void)
 {
    int done = 0;     // Set to true when Y or N is encountered
    char letter;
```

```
   while (! done)
    {
      cout << "\nType Y or N and press Enter to continue: ";
      cin >> letter;

      if ((letter == 'Y') || (letter == 'y'))
        done = 1;
      else if ((letter == 'N') || (letter == 'n'))
        done = 1;
      else
        cout << '\a';      // Sound the speaker's bell for
                           // invalid character
    }
  cout << "The letter you typed was " << letter << endl;
}
```

As you can see, the *while* loop also supports multiple statements grouped within left and right braces. In this example, the program uses the variable *done* to control the loop. As long as the program is not finished (the user has not typed Y or N), the loop continues. When the user types Y or N, the program sets the variable *done* to true and the loop ends. As your programs begin to work with files, you will use *while* loops on a regular basis.

Looping Until a Specific Condition Occurs

As your programs become more complex, there will be times when you must perform a group of related statements until a specific conditions occurs. For example, your program might calculate payroll amounts for all the employees in a company. To do so, the program would loop until it has processed the last employee. To repeat statements until a specific condition occurs, your programs will normally use a *while* statement:

```
while (condition_is_true)
   statement;
```

When your program encounters a *while* statement, your program will evaluate the loop's condition. If the condition is true, your program will execute the *while* loop's statements. After the program performs the last statement within the loop, the program tests the condition again. If the condition is true, the program will repeat this process, executing the statements and then repeating the condition. When the condition evaluates as false, the program will continue its execution at the first statement that follows the *while* statement.

PERFORMING STATEMENTS AT LEAST ONE TIME

As you have just learned, the C++ *while* loop lets your programs repeat a set of statements while a specific condition is true. When your program encounters a *while* statement, the program first evaluates the condition specified. If the condition is true, the program will enter the loop. If the condition is false, the program will never execute the *while* loop's statements. Depending on your program's purpose, there will be many times when you will want the program to perform a set of statements at least once, and then, based on some condition, possibly repeat. In such cases, your programs can use the *do while* loop, as shown here:

```
do {
    statements;
} while (condition_is_true);
```

When your program encounters a *do while* loop, the program enters the loop and starts executing the statements the loop contains. After the program executes the last statement within the loop, the program then evaluates the specified condition. If the condition is true, the program will loop back to the start of the loop, as shown here:

```
do {  ←
    statements;
} while (condition_is_true); ——
```

If the condition is false, the program will not repeat the loop's instructions, and instead, will continue its execution with the first statement that follows the loop. A common use of the *do while* loop is to display menu options and then process the user's selection. You will want your program to display the menu at least one time. If the user selects any option except Quit, the program will perform the option and then redisplay the menu (repeating the loop's statement). If the user selects Quit, the loop will end, and the program will continue its processing at the first statement after the loop.

Repeating Statements Provided a Condition Is True

Depending on your program's requirements, there may be times when your program must perform a set of statements at least one time and then possibly repeat the statements if a specific condition is true. In such cases, your programs should use the C++ *do while* statement, as shown here:

```
do {
    statement;
} while (condition_is_true);
```

When your program encounters a *do while* statement, your program immediately performs the statements the loop contains. Next, the program examines the loop's condition. If the condition is true, the program will repeat the loop's statements and the process will continue. When the loop's condition becomes false, the program will continue its execution at the first statement that follows the *do while* statement.

WHAT YOU MUST KNOW

Iterative processing is the ability of a program to repeat one or more statements. This lesson presented the C++ iterative (or looping) statements. As you have learned, the *for* statement lets your programs repeat one or more statements a specific number of times. The *while* statement lets your programs repeat statements as long as a specific condition is true. Finally, the *do while* statement lets your programs perform statements at least one time, possibly repeating if a specified condition is true. In Lesson 10, "Getting Started with Functions," you will learn how to break larger programs into smaller, more manageable pieces, called functions. Before you continue with Lesson 10, however, make sure you have learned the following key concepts:

☑ The C++ *for* statement lets your programs repeat one or more statements a specific number of times.

☑ The *for* statement consists of four parts: an initialization, a test condition, the statements that are to repeat, and an increment.

☑ The *for* statement does not require your programs to increment the loop's control variable by one, nor does the *for* statement require your programs to count upward.

☑ The C++ *while* loop lets your programs repeat statements as long as a specific condition is true.

☑ Programs often use the *while* loop to read the contents of a file, looping until the program encounters the end of the file.

☑ The C++ *do while* statement lets your programs perform one or more statements at least once, possibly repeating based on a specific condition.

☑ Programs often use *do while* statements for menu processing.

☑ When a *for*, *while*, or *do while* loop's condition becomes false, the program will continue its execution at the first statement that follows the loop.

MICROSOFT VISUAL C++

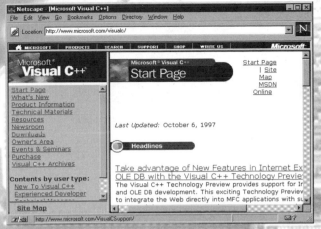

http://www.microsoft.com/visualc/

CROSS-PLATFORM RESOURCES

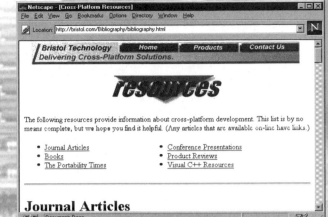

http://bristol.com/Bibliography/bibliography.html

PROGRAMMING GUIDES & RESOURCES

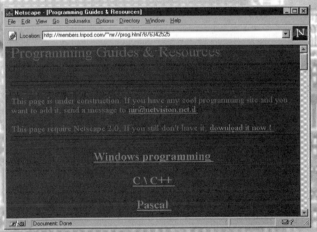

http://members.tripod.com/~nir7/prog.html?876342525

WWW VIRTUAL LIBRARY

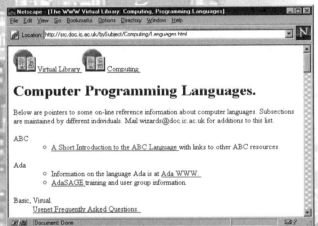

http://src.doc.ic.ac.uk/bySubject/Computing/Languages.html

TURBO C++ 4.5

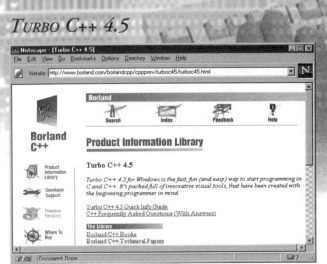

http://www.borland.com/borlandcpp/cppprev/—
turboc45/turboc45.html

STANDARD TEMPLATE LIBRARY

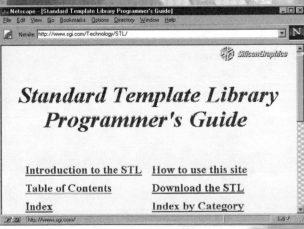

http://www.sgi.com/Technology/STL/

THE GUI TOOLKIT

http://home.pacbell.net/atai/guitool/

OBJECT-ORIENTED INFORMATION SOURCES

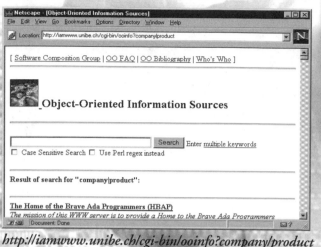

http://iamwww.unibe.ch/cgi-bin/ooinfo?company/product

C/C++ USERS JOURNAL

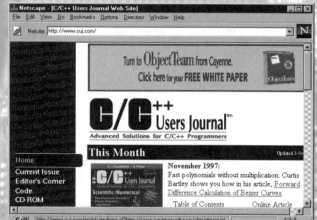

http://www.cuj.com/

OTHER C/C++ RESOURCES

http://www.r2m.com/windev/cpp-compiler.html

THE GOOD PROJECT

http://metallica.prakinf.tu-ilmenau.de/GOOD.html

DEVELOPER RESOURCES

http://www.genitor.com/resources.htm

BUILDING PROGRAMS USING FUNCTIONS

As your programs become larger and more complex, you will break them up into smaller, more manageable pieces called *functions*. Each function within your program will perform a specific task. Assume, for example, that you are creating an accounting program. You might have one function for accounts receivable processing, one for accounts payable, one for payroll, and so on. By focusing on one function at a time, your programs become easier for you to create and understand. In addition, you will find that you can use many of the functions you create for one program in another, which saves you programming time. The lessons in this section include:

LESSON 10

GETTING STARTED WITH FUNCTIONS

As your programs increase in size and complexity, you should break them up into smaller more manageable pieces called *functions*. Each function within your program should perform a specific task. For example, if you are writing a payroll program, you might create one function that determines the number of hours an employee worked, a second function to determine overtime pay, a third to print the payroll checks, and so on. When your programs must perform a specific task, the program *calls* the corresponding function, passing to the function the information it needs to perform its processing, such as an employee name or hourly pay rate. This lesson will teach you how to create and use functions within your C++ programs. By the time you finish this lesson, you will understand the following key concepts:

- *Functions* group related statements to perform a specific task.

- To use a function, your program *calls* the function by referencing the function name, followed by parentheses, such as *beep()*.

- When they complete their processing, many functions return a value of a specific type, such as *int* or *float*, that your programs can test or assign to a variable.

- Your programs pass parameters (information) to functions, such as an employee name, age, or salary, by including the parameters within the parentheses that follow the function name.

- C++ uses function *prototypes* to define the type of value a function returns to the program, as well as the number and type of parameters the program passes to the function.

As your programs become larger and more powerful, your use of functions will become essential. As you will find, creating and using functions in C++ is very easy.

CREATING AND USING YOUR FIRST FUNCTIONS

As your create your programs, you should design each function to perform a specific task. If you find that a function performs more than one task, you should divide the function into two or more functions. You should assign each function you create within your programs a unique name. As is the case with variable names, the function names you choose should correspond to the operation the function performs. For example, by simply looking at the function names listed in Table 10, you have a good idea about each function's purpose.

Function Name	Function Purpose
print_test_scores	Print the class test scores
accounts_payable	Process a company's accounts payable
get_user_name	Prompt the user for his or her name
print_document	Print the specified document file
calculate_income_tax	Determine a user's income tax

Table 10 Examples of meaningful function names.

A C++ function is similar in structure to the *main* program you have been using in all the previous programs. In other words, you proceed the function name with a type, such as *int* or *void*, and follow the name with a parameter list that appears within parentheses. Also, like *main*, you group the function's statements within left and right braces, as shown here:

```
return_type  function_name(parameter_list)
  {
       variable_declarations;

       statements;
  }
```

Consider, for example, how this function structure corresponds to the following *main* program:

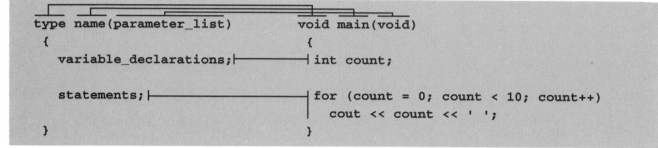

```
type name(parameter_list)              void main(void)
  {                                       {
     variable_declarations;                  int count;

     statements;                             for (count = 0; count < 10; count++)
                                               cout << count << ' ';
  }                                       }
```

The following statements define a function named *show_message* that displays a message to the screen using *cout*:

```
void show_message(void)
  {
     cout << "Hello, I've been Rescued by C++" << endl;
  }
```

As you might remember from Lesson 3, "Taking a Closer Look at C++," the word *void* that precedes the function name specifies (to the C++ compiler and to programmers who read your code) that the function does not return a value to its caller. Likewise, the word *void* contained within the parentheses specifies the function does not use *parameters* (information the program passes to the function). The following program, *Show_Msg.CPP*, uses the *show_message* function to display a message on the screen:

```
#include <iostream.h>

void show_message(void)
  {
    cout << "Hello, I've been Rescued by C++" << endl;
  }

void main(void)
  {
    cout << "About to call the function" << endl;
    show_message();
    cout << "Back from the function" << endl;
  }
```

As you have learned, your program's execution always begins within *main*. In this case, within *main* the following statement, the *function call*, invokes the function *show_message*:

```
show_message();
```

The parentheses that follow the function name tell the C++ compiler that your program is using a function. Later in this lesson, you will learn that your programs can pass information (parameters) to functions within the parentheses. When you compile and execute the *Show_Msg.CPP* program, your screen will display the following output:

```
C:\> Show_Msg  <ENTER>
About to call the function
Hello, I've been Rescued by C++
Back from the function
```

When the program encounters the function call, the program begins executing the statements that reside within the function. After the program executes all the statements the function contains (in other words, the function completes its processing), the program's execution continues at the statement that immediately follows the function call, as shown here:

```cpp
#include <iostream.h>

void show_message(void)
  {
    cout << "Hello, I've been Rescued by C++" << endl;
  }

void main(void)
  {
    cout << "About to call the function" << endl;
    show_message();
    cout << "Back from the function" << endl;
  }
```

In this example, the program executes the first statement in *main* that displays a message to the user that the program is about to call the function. Next, the program encounters the function call and starts executing the statements that reside in the *show_messsage* function. After the program executes the function's only statement, the program continues its execution back in *main*, at the statement that immediately follows the function call. In this case, the program displays a message telling the user that it has returned from the function call and then ends. The following program, *Two_Msgs.CPP*, uses two functions, *show_title* and *show_lessons*, to display information about this book:

```cpp
#include <iostream.h>

void show_title(void)
  {
    cout << "Book: Rescued by C++" << endl;
  }

void show_lesson(void)
  {
    cout << "Lesson: Getting Started With Functions" << endl;
  }

void main(void)
  {
    show_title();
    show_lesson();
  }
```

When the program's execution begins, the program will first call the *show_title* function, which displays a message using *cout*. When *show_title* ends, the program then calls the *show_lesson* function, which also displays a message. When *show_lesson* ends, there are no more statements in *main*, so the program ends.

So far, the functions this lesson's programs presented have performed very simple tasks. In each case, your program could have easily performed the same processing without using the functions, by simply including the functions' statements within *main*. However, the functions' purpose was to show you how your program defines and later calls a function. As your programs become more complex, you will use functions to simplify large tasks by breaking your program into smaller, more manageable pieces.

As you create functions, you will find that because they contain fewer lines of code than one large program, your functions are easy to understand and easy to change. In addition, you will find that, in many cases, you can use a function that you create for one program within a second program, with no changes. By creating a library of functions, you will reduce the amount of time you spend in the future coding and testing similar functions.

Calling a Function

A function is a collection of related statements that perform a specific task. By creating functions within your programs, you can break large tasks into smaller, more manageable pieces. Your programs execute a function's statements by calling the function. To call the function, your program simply references the function's name followed by parentheses, as shown here:

```
function_name();
```

When your program passes information (parameters) to the function, your program will place the information within the parentheses, separated by commas, as shown here:

```
payroll(employee_name, employee_id, salary);
```

After the last statement in the function completes, the program's execution continues at the first statement that follows the function call.

PROGRAMS CAN PASS INFORMATION TO FUNCTIONS

To increase your function's capabilities, C++ lets your programs pass information (parameters) to your functions. When a function uses parameters, you must tell C++ each parameter's type, such as *int*, *float*, *char*, and so on. For example, the following function, *show_number*, uses a parameter of type *int*:

```
void show_number(int value)
  {
    cout << "The parameter's value is " << value << endl;
  }
```

When your program invokes the *show_number* function, your program must pass a value to the function, as shown here:

```
show_number(1001);                          — Value passed to the function
```

Behind the scenes, C++ will substitute the value your program passes to the function for each occurrence of the parameter's name value within the function, as shown here:

```
show_number(1001);


        void show_number(int value)
          {
             cout << "The parameter's value is " << value << endl;
          }


                void show_number(1001)
                  {
                     cout << "The parameter's value is " << 1001 << endl;
                  }
```

As you can see, because C++ substitutes the value for the parameter name, the function *show_number* displays the value 1001, which *main* passed to it during the function call.

The following program, *UseParam.CPP*, uses the *show_number* function several times, each time passing a different value:

```cpp
#include <iostream.h>

void show_number(int value)
  {
    cout << "The parameter's value is " << value << endl;
  }

void main(void)
  {
    show_number(1);
    show_number(1001);
    show_number(-532);
  }
```

When you compile and execute the *UseParam.CPP* program, your screen will display the following:

```
C:\> UseParam  <ENTER>
The parameter's value is 1
The parameter's value is 1001
The parameter's value is -532
```

As you can see, each time the program invokes the function, C++ assigns the correct number to the variable value. Take time to experiment with this program, changing the values *main* passes to the function, and note the result.

As you have learned, for a program to pass information to a function, the function must support one or more parameters. Each function parameter is a specific type. In the case of the *show_number* function, the parameter value must be type *int*. If you try to pass a value of a different type to the function, such as a floating-point value, the compiler will generate a syntax error. In most cases, your programs will pass several values to a function. For each parameter, the function must specify a corresponding name and type. For example, the following program, *BigSmall.CPP*, uses the function *show_big_and_little* to display the largest and smallest of the three integer values it receives:

```
#include <iostream.h>

void show_big_and_little(int a, int b, int c)
 {
   int small = a;
   int big = a;

   if (b > big)
     big = b;
   if (b < small)
     small = b;
   if (c > big)
     big = c;
   if (c < small)
     small = c;

   cout << "The biggest value is " << big << endl;
   cout << "The smallest value is " << small << endl;
 }

void main(void)
 {
   show_big_and_little(1, 2, 3);
   show_big_and_little(500, 0, -500);
   show_big_and_little(1001, 1001, 1001);
 }
```

When the program calls the function, C++ assigns parameters, as shown here:

```
show_big_and_little(1, 2, 3);



            void show_big_and_little(int a, int b, int c);
```

When you compile and execute the *BigSmall.CPP* program, your screen will display the following output:

```
C:\> BigSmall   <ENTER>
The biggest value is 3
The smallest value is 1
The biggest value is 500
The smallest value is -500
The biggest value is 1001
The smallest value is 1001
```

In the following example, the program *Show_Emp.CPP* uses the function *show_employee* to display an employee's age (type *int*) and salary (type *float*):

```
#include <iostream.h>

void show_employee(int age, float salary)
 {
```

```
     cout << "The employee is " << age << " years old" << endl;
     cout << "The employee makes $" << salary << endl;
  }


void main(void)
  {
    show_employee(32, 25000.00);
  }
```

As you can see, the function *show_employee* defines the parameters as type *int* and *float* as required.

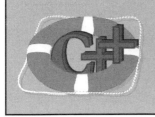

Passing Parameters to Functions

When your functions use parameters, your functions must specify a unique name and type for each parameter. When your program invokes the function, C++ will assign the parameter values to the function's parameter names from left to right. Each function parameter has a specific type, such as *int*, *float*, or *char*. The values your program passes to a function using parameters must match the parameters' type.

FUNCTIONS CAN RETURN A RESULT TO THE CALLER

As discussed, your program's functions should perform a specific task. In many cases, a function will perform some type of calculation. The function will then return its result to the caller (the location in the program that called the function). When a function returns a value, you must tell C++ the value's type, such *int*, *float*, *char* and so on.

To inform C++ about the function's return type, you simply precede the function name with the corresponding type. For example, the following function, *add_values*, adds its two integer parameters and returns a result of type *int* to the calling program:

```
int add_values(int a, int b)
  {
    int result;

    result = a + b;

    return(result);
  }
```

In this case, the word *int* that appears before the function name specifies the function's *return type*. Functions use the *return* statement to return a value to their caller. When a function encounters the *return* statement, the function returns the value specified to the caller and the function's execution ends, returning control to the caller. The following statement, for example, assigns the value the *add_values* function returns to the variable *result*:

```
result = add_values(1 + 2);
```

In this example, the program assigns the function's return value to a variable. Your program can also directly print the function's return value using *cout*, as shown here:

```
cout << "Sum of values is " << add_values(500, 501) << endl;
```

The previous implementation of the *add_values* function used three statements to make the function's processing easier to understand. You can, however, reduce the function to a *return* statement, as shown here:

```
int add_values(int a, int b)
 {
   return(a+b);
 }
```

The following program, *AddValue.CPP*, uses the *add_values* function to add several different values:

```
#include <iostream.h>

int add_values(int a, int b)
 {
   return(a+b);
 }

void main(void)
 {
   cout << "100 + 200 = " << add_values(100, 200) << endl;
   cout << "500 + 501 = " << add_values(500, 501) << endl;
   cout << "-1 + 1 = " << add_values(-1, 1) << endl;
 }
```

Take time to experiment with the *AddValue.CPP* program, changing the values the program passes to the function. You might try passing large values to the function, such as 20000 and 30000. As you might guess, the function (which returns a value of type *int*) will experience an overflow error and return an errant result.

Not all functions will return a value of type *int*. The following function, *average_value*, returns the average of two integer values, which can be a fractional value, such as 3.5:

```
float average_value(int a, int b)
 {
   return((a + b) / 2.0);
 }
```

In this case the word *float*, which precedes the function name, specifies the function's return type. The following program, *GetAve.CPP*, uses the *average_value* function to display the average of the values 5 and 10:

```
#include <iostream.h>

float average_value(int a, int b)
 {
   return((a + b) / 2.0);
 }

void main(void)
 {
   cout << "The average value is: " << average_value(5, 10) << endl;
 }
```

FUNCTIONS THAT DO NOT RETURN VALUES

As you will learn, not all functions return a result. For example, earlier in this lesson, you used the *show_message* function to display a message using *cout*. If a function does not return a value, you must precede the function name with the type *void*. As you have learned, to return a value to the caller, functions use the *return* statement. When the function encounters a *return* statement, the function's execution ends, and C++ returns the value specified to the caller. As you examine C++ programs, there may be times when you will encounter a *return* statement within a function that does not return a value, as shown here:

```
return ;
```

In this example, the function is of type *void* (does not return a value) and the *return* statement simply ends the function's execution.

Note: *If the function has statements that appear after the **return** statement, the statements will not execute. As discussed, when the function encounters a **return** statement, the function will return the specified value to its caller and the program's execution will continue at the first statement that follows the function call.*

USING A FUNCTION'S RETURN VALUE

When a function returns a value, the caller can assign the return value to a variable using the assignment statement as shown here:

```
payroll_amount = payroll(employee, hours, salary);
```

In addition, the caller can simply reference the function name. For example, the following statement displays the function's return value using *cout*:

```
cout << "The employee made " << payroll(employee, hours, salary) << end;
```

Also, the caller can use the function's return value within a condition, as shown here:

```
if (payroll(employee, hours, salary) << 500.00)
    cout << "This employee needs a raise" << endl;
```

As you can see, a program can use a function's return value in many different ways.

UNDERSTANDING FUNCTION PROTOTYPES

Before your program can call a function, C++ must know the type of value the function returns and the number and type of parameters the function uses. In each of this lesson's programs, the definition of the function your program is to call has preceded the function call within the source file. In many cases, however, functions will appear throughout your source file, and it is common for one function to call another.

To ensure that C++ knows specifics about each function your program uses, you should place *function prototypes* near the start of your source file. In general, a function prototype provides C++ (as well as another programmer who is reading your program) with information about the function's return type and parameters. The following statements illustrate function prototypes for several of the functions used throughout this lesson:

```
void show_message(void);

void show_number(int);

void show_employee(int, float);

int add_values(int, int);

float average_value(int, int);
```

As you can see, a function prototype specifies the function's return type and the number and type of each parameter. Note the semicolon that ends each prototype:

```
float average_value(int, int);                          Return type
                                                        Parameter types
```

If your program tries to call a function for which the C++ compiler has not encountered the function definition or a prototype, the compiler will generate a syntax error. As you examine C++ header files and other programs, you will encounter function prototypes on a regular basis. The following program, *Proto.CPP*, illustrates how to use a function prototype:

```
#include <iostream.h>

float average_value(int, int);   // Function prototype

void main(void)
  {
    cout << "The average of 2000 and 2 is " <<
       average_value(2000, 2) << endl;
  }

float average_value(int a, int b)
  {
    return((a + b) / 2.0);
  }
```

In this case the program invokes the function *average_value* prior to the function definition. Therefore, the program uses the function prototype, which precedes the definition of *main*. If you remove the function prototype and compile the *Proto.CPP* program, the C++ compiler will generate syntax errors.

Function Prototypes Exist to Help You

A function prototype tells the C++ compiler a function's return type and the number and type of the function's parameters. When you compile your program, the C++ compiler uses each function prototype to ensure that you do not misuse a function's return value (such as assigning a return value of type *float* to a variable of type *int*), and that you do not pass the wrong type of value for a parameter. In the past, many C compilers did not perform such type checking. As a result, programmers often spent hours trying to debug errors that occurred because they passed a value of type *int* to a function that expected a value of type *float*. When you encounter a syntax error due to a conflict with a function prototype, be thankful. In the past, the compiler would not have detected the error and your program would simply not work.

WHAT YOU MUST KNOW

In this lesson you learned how to use functions within your C++ programs. This lesson covered a lot of key material, such as parameters, return types, and function prototypes. You might want to spend a few more minutes now experimenting with this lesson's sample programs. In Lesson 11, "Changing Parameter Values," you will learn how to change the values of parameters within your functions. Before you continue with Lesson 11, however, make sure that you have learned the following key concepts:

☑ As your programs become larger and more complex, you should break your programs into smaller more manageable pieces, called functions. Each function should have a unique name. Assign names to your functions that meaningfully describe the task the function performs.

☑ Functions can return a value to the caller. If a function returns a value, you must specify the function's return type (*int, char*, and so on) before the function name; otherwise, you should precede the function name with *void*.

☑ Programs pass information to functions using parameters. If the function receives parameters, you must specify both a unique name for each parameter and a type. If the function does not receive parameters, you should place the *void* keyword between the parentheses that follow the function name.

☑ C++ must know a function's return type and the number and type of parameters the function receives. If the function definition follows the function's use, you must place a function prototype at the start of your source file.

LESSON 11

CHANGING PARAMETER VALUES

In Lesson 10, "Getting Started with Functions," you learned how to break your programs into smaller, more manageable pieces called functions. As you learned, your programs can pass information (parameters) to functions. In Lesson 10, the functions used or displayed, but did not change, a parameter's values. In this lesson, you will learn how to change a parameter's value within a function. As you will find, to change a parameter within a function actually requires more steps than you might imagine. However, this lesson will teach you all the steps you must know. By the time you finish this lesson, you will understand the following key concepts:

- Unless a function uses pointers or a C++ reference, the function cannot change a parameter's value.

- To change a parameter's value, a function must know the parameter's address in memory.

- The C++ address operator (&) lets your program determine a variable's memory address.

- When your programs know a memory address, your programs can use the C++ memory indirection operator (*) to determine the value stored at that address.

- When a function must change a parameter's value, the program must pass the parameter's address to the function.

As you will learn, changing a parameter's value within a function is a common operation. Experiment with the programs this lesson presents to ensure that you are comfortable with the process.

WHY FUNCTIONS NORMALLY CANNOT CHANGE PARAMETER VALUES

The following program, *NoChange.CPP*, passes two parameters, named *big* and *small*, to the function *display_values*. The function *display_values*, in turn, assigns both parameters the value 1001 and then displays each parameter's value. When the function ends, the program resumes and displays each parameter's value:

```
#include <iostream.h>

void display_values(int a, int b)
 {
   a = 1001;
   b = 1001;

   cout << "The values within display_values are " <<  a <<
     " and " << b << endl;
 }

void main(void)
 {
   int big = 2002, small = 0;

   cout << "Values before function " << big << " and "  <<
     small << endl;
```

```
    display_values(big, small);

    cout << "Values after function " << big << " and " << small << endl;
}
```

When you compile and execute the *NoChange.CPP* program, your screen will display the following output:

```
C:\> NoChange   <ENTER>
Values before function 2002 and 0
The values within display_values are 1001 and 1001
Values after function 2002 and 0
```

As you can see, the *display_values* function changes each parameter's value to 1001. However, when the function ends, the values of the variables *big* and *small* within *main* have not changed. To understand why the function's value assignment did not affect the variables *big* and *small* within *main*, you must understand how C++ passes parameters to functions.

By default, when your programs pass a parameter to a function, C++ makes a copy of the parameter's value and places the copy into a temporary memory location called the *stack*. The function then uses the *copy* of the value to perform its operations. When the function ends, C++ discards the stack contents and any changes the function has made to the copies of the parameter values.

As you know, a variable is a name your program assigns to a location in memory that stores a value of a specific type. Assume, for example, that the variables *big* and *small* reside at memory locations 10 and 12. When you pass the variables to the function *display_values*, C++ will place copies of the variable's values on the stack. As Figure 11.1 shows, the function *display_values* will then use the copies of the variable's values.

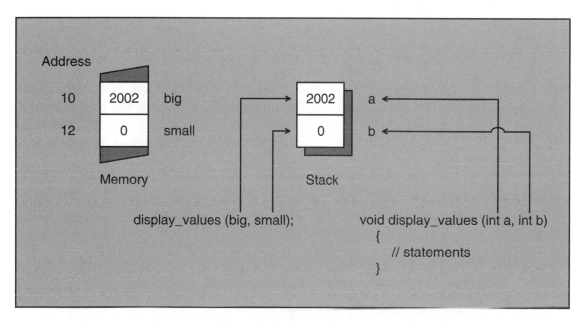

Figure 11.1 C++ places copies of parameter values in a temporary storage location called the stack.

As you can see, the *display_values* function has access to the stack contents that contains copies of the values 2002 and 0. Because the *display_values* function has no knowledge of the *big* and *small* variables' memory locations (address 10 and 12), the function has no way to change the variables' actual values.

Why C++ Functions Normally Cannot Change Parameter Values

When your program passes parameters to a function, C++ places copies of the parameter values in a temporary storage location called the stack. Any changes the function makes to the parameters affect only the variable copies within the stack. When the function ends, C++ discards the stack contents along with any changes the function has made to the parameter values. Because the function has no knowledge of the parameter's memory address, the functions cannot change the values the parameters store.

CHANGING A PARAMETER'S VALUE

To change a parameter's value, a function must know the parameter's memory address. To tell a function a parameter's address, your programs must use the C++ address operator (&). The following function call illustrates how your program will use the address operator to pass the addresses of the variables *big* and *small* to the function *change_values*:

```
change_values(&big, &small);
```
Passing parameters by address

Within the function definition, you must tell C++ that the program will pass parameters to the function by address. To do so, you declare *pointer variables*, by preceding each variable name with an asterisk, as shown here:

```
void change_values(int *big, int *small)
```
Pointer to the type int

A pointer variable is a variable that contains a memory address. To change the parameter value within the function, you precede the parameter name with an asterisk, as shown here:

```
*big = 1001;
*small = 1001;
```

The following program, *ChgParam.CPP*, uses the address operator (&) to pass the addresses of the parameters *big* and *small* to the *change_values* function. The function, in turn, uses pointers to the parameters' memory locations. Therefore, the changes the function makes to the parameters remain after the function ends, as shown here:

```
#include <iostream.h>

void change_values(int *a, int *b)
 {
   *a = 1001;
   *b = 1001;

   cout << "The values within display_values are " << *a <<
     " and " << *b << endl;
 }

void main(void)
 {
   int big = 2002, small = 0;

   cout << "Values before function " << big << " and " <<
     small << endl;
```

```
    change_values(&big, &small);

    cout << "Values after function " << big << " and " <<
      small << endl;
}
```

When you compile and execute the *ChgParam.CPP* program, your screen will display the following output:

```
C:\> ChgParam    <ENTER>
Values before function 2002 and 0
The values within display_values are 1001 and 1001
Values after function 1001 and 1001
```

As you can see, the values the *change_values* function assigns to parameters remain in effect after the function ends. To understand why the changes the function makes to the variables remain in effect, you must remember that the function has access to each variable's memory locations. When you pass the parameters to a function by address, C++ places each variable's address on the stack, as shown in Figure 11.2.

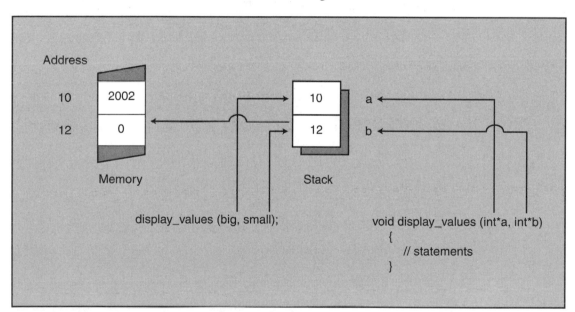

Figure 11.2 *Passing parameters to a function by address.*

Using pointers (the memory address) within the function, *change_values* can access each parameter's address in memory, changing the values as desired.

Changing Parameter Values Within Functions

To change a parameter's value within a function, the function must have knowledge of the parameter's address in memory. Therefore, your programs must pass the parameter's address, using the C++ address operator:

```
some_function(&some_variable);
```

Within the function, you must tell C++ that the function will be working with a memory address (a pointer). To do so, you precede the parameter name within the declaration with an asterisk:

89

```
void some_function(int *some_variable);
```

Next, to change the parameter's value within the function, you must precede references to the pointer variable's name with an asterisk, as shown here:

```
*some_variable = 1001;

cout << *some_variable;
```

To prevent errors, C++ will not let your program pass a variable's address to a function that does not expect a pointer as a parameter. Likewise, C++ will normally generate a compiler warning when your program tries to pass a value to a function that expects a pointer as its parameter.

A SECOND EXAMPLE

When your programs pass pointers to parameters, the parameters can be of any type, such as *int*, *float*, or *char*. The function that uses the pointers declares variables of the corresponding type, preceding each variable name with an asterisk to indicate that the variable is a pointer. The following program, *SwapVals.CPP*, passes the addresses of two parameters of type *float* to the *swap_values* function. The function, in turn, uses pointers to each parameter's memory location to swap the parameter's values:

```
#include <iostream.h>

void swap_values(float *a, float *b)
  {
    float temp;

    temp = *a;
    *a = *b;
    *b = temp;
  }

void main(void)
  {
    float big = 10000.0;
    float small = 0.00001;

    swap_values(&big, &small);

    cout << "Big contains " << big << endl;
    cout << "Small contains " << small << endl;
  }
```

As you can see, the program passes the parameters to the function *swap_values* by address. Within the function, the statements use pointers to each parameter's memory address. It is worthwhile to take a close look at the processing that occurs within the function *swap_values*. As you can see, the function declares *a* and *b* as pointers to values of type *float*:

```
void swap_values(float *a, float *b)
```

The function declares the variable *temp*, however, as simply a *float*, not a pointer to *float*:

```
float temp;
```

Consider the following statement:

```
temp = *a;
```

The statement directs C++ to assign the value to which the pointer variable *a* currently points (which is *big's* value 10000.0) to *temp*. Because *temp* is type *float*, the assignment is correct. A pointer variable is a variable that holds an address. The following statement would declare *temp* as a pointer to a memory location holding a value of type *float*:

```
float *temp;
```

In this example, *temp* could store the address of a floating-point value but not the value itself.

If you remove the indirection operator (*) from in front of the variable *a* within the assignment, the statement would try to assign the value stored in *a*, which is an address, to the variable *temp*. Because *temp* can hold a floating-point value, but not the address of a floating-point value, an error would occur.

If you do not yet feel comfortable with pointers, don't worry, Section 3 will discuss them in more detail. For now, however, simply know that if you want your functions to change the value of a parameter, you must use pointers.

Use Assembly Language Listings to Understand the Compiler's Processing

One of the best ways to understand how the C++ compiler treats pointers is to examine the compiler's assembly-language output. Most C++ compilers provide a command-line switch you can use when you compile the program that directs the compiler to produce an assembly language listing. By reading the assembly-language listing, you may better understand how the compiler uses the stack when it passes parameters to functions.

WHAT YOU MUST KNOW

In this lesson you learned how to change a parameter's value within a function. To change a parameter's value, your functions must use pointers. At first, you may find pointers intimidating. In Lesson 15, "Using C++ References," you will learn to use C++ *references*, which simplify the process of changing parameters within a function. However, because many C programmers use pointers to change parameters, you must be aware of the processing involved. In Lesson 12, "Taking Advantage of the Run-Time Library," you will learn that most C++ compilers provide library functions that can save you considerable programming while letting you develop powerful programs quickly. Before you continue with Lesson 12, however, make sure that you have learned the following key concepts:

☑ Unless a function uses pointers or C++ references, the function cannot change a parameter's value.

☑ When your program passes a parameter to a function, C++ places a copy of the parameter's value in a temporary storage location called the stack. Any changes the function makes to the parameter affect only the copy of the value that resides on the stack.

☑ To change a parameter's value, a function must know the corresponding variable's memory address.

☑ Using the C++ address operator (&), programs can pass a variable's address to a function.

☑ When a function receives a variable's address, the function must declare the parameter variables as pointers (by preceding the variable names with an asterisk).

☑ When the function must use the value referenced (pointed to) by a pointer, the function must precede the pointer variable's name with an asterisk (*), the C++ indirection operator.

LESSON 12

TAKING ADVANTAGE OF THE RUN-TIME LIBRARY

In Lesson 10, "Getting Started with Functions," you learned that you can divide your programs into small, manageable pieces, called functions, that perform a specific task. One advantage of using functions is that you can often use a function you create for one program within a second program. As you will learn in this lesson, most C++ compilers provide an extensive collection of functions your programs can use, called the *run-time library*. By taking advantage of these functions, you reduce the amount of programming you must perform yourself. Instead, your program simply calls the run-time library function. Depending on your compiler, the run-time library may consist of thousands of functions. This lesson examines how you use these functions within your programs. By the time you finish this lesson, you will understand the following key concepts:

- The *run-time library* is a collection of functions your compiler provides that you can easily use within your programs.

- To use a run-time library function, you must include a header file from the run-time library that defines a function's prototype.

- Some compilers refer to the run-time library as an *application program interface*, or *API*.

Most run-time libraries contain hundreds of usable functions that can save you a tremendous amount of programming time and can help you quickly develop powerful programs. As you will learn, using the run-time library functions is very easy!

USING A RUN-TIME LIBRARY FUNCTION

In Lesson 10 you learned that before your programs can invoke a function, the C++ compiler must have encountered that function's definition or prototype. Because the run-time library functions are not defined within your program (the C++ compiler defines the functions in a library file), you must specify a function prototype for each run-time library function you plan to use. To make it easy for you to use run-time library functions, the C++ compiler provides header files that contain the correct function prototypes. Therefore, your programs must simply include the correct header file, using the *#include* statement, and then call the function they want. For example, the following program, *ShowTime.CPP*, will use the run-time library functions *time* and *ctime* to display the current system date and time. The header file *time.h* contains the prototypes for these two run-time library functions, as shown here:

```
#include <iostream.h>
#include <time.h>         // For run-time library functions

void main(void)
 {
   time_t system_time;

   system_time = time(NULL);

   cout << "The current system time is " <<
     ctime(&system_time) << endl;
 }
```

When you compile and execute the *ShowTime.CPP* program, your screen will display the current date and time as shown here:

```
C:\> ShowTime    <ENTER>
The current system time is Sat Oct 25 16:13:51 1997
```

As you can see, the program uses the functions *time* and *ctime*. In the case of the function *ctime*, the program passes the address of the variable *system_time*, using the address operator discussed in Lesson 11, "Changing Parameter Values." To use these run-time library functions, you had to simply include the header file *time.h* near the top of your source file.

In a similar way, the following program, *Sqrt.CPP*, uses the *sqrt* function to return the square root of several different values. The function prototype for *sqrt* resides in the header file *math.h*, as shown here:

```cpp
#include <iostream.h>
#include <math.h>            // Contains sqrt prototype

void main(void)
 {
   cout << "The square root of 100.0 is " << sqrt(100.0) << endl;
   cout << "The square root of 10.0 is " << sqrt(10.0) << endl;
   cout << "The square root of 5. 0 is " << sqrt(5.0) << endl;
 }
```

As a final example, the program *SysCall.CPP* uses the *system* function, whose prototype the header file *stdlib.h* defines. The *system* function provides an easy way for your program to run an operating-system command, such as DIR, or another program:

```cpp
#include <stdlib.h>

void main(void)
 {
   system("DIR");
 }
```

In this example, the program uses the *system* function to invoke the MS-DOS DIR command. Take time to experiment with this program and execute other commands, or even one of the programs you created earlier in this book.

LEARNING ABOUT RUN-TIME LIBRARY FUNCTIONS

Your C++ compiler provides hundreds of functions within its run-time library. The documentation that accompanies your compiler should contain a complete description of the available run-time library functions. As you examine this documentation, you will find that the document normally presents the functions using a sample prototype. For example, for the *sqrt* function you might find the following function prototype:

```cpp
double sqrt(double);
```

In this example, the function prototype tells you that the function returns a value of type *double* and expects a parameter of type *double* as well. Likewise, you might find the following prototype for the *time* function:

```cpp
time_t time(time_t *);
```

Again, the prototype tells you that the function returns a value of type *time_t* (which the header file *time.h* defines). The function expects its parameter to be a pointer to a variable of type *time_t*. As you read about run-time library functions, you will learn a great deal about the functions and C++ by examining the function prototypes.

Another way to learn about your compiler's run-time library functions is to examine the header files that reside in the *INCLUDE* subdirectory beneath your compiler. Take time now to print the header files *math.h*, *ctime.h*, and *stdlib.h*, which you have used for the programs in this lesson.

Using API Functions

In addition to its standard run-time library, many compilers support *API*, or *application program interface* functions. If you are programming in the Windows environment, for example, there are graphics API functions, telephony API functions called the TAPI, multimedia API functions, and much more. Before you create your own functions, make sure you find out which API functions are readily available for your compiler.

What You Must Know

The C++ run-time library provides a powerful collection of functions your programs can use. Spend time now to locate the run-time library documentation that accompanied your compiler. Get a feel for the functions the run-time library provides. By taking advantage of these functions, you eliminate considerable programming. In Lesson 13, "Local Variables and Scope," you will learn about local variables and scope (the locations in your program where a variable's name is known). Before you continue with Lesson 13, however, make sure that you have learned the following key concepts:

- ☑ The run-time library is a set of functions the compiler provides for your programs' use.

- ☑ To use a run-time library function, you must specify a function prototype.

- ☑ Most C++ compilers provide header files that contain the correct function prototypes for each run-time library function.

- ☑ In addition to providing a run-time library, many C++ compilers support API (application program interface) functions that support specific tasks, such as graphics or multimedia programming.

LESSON 13

LOCAL VARIABLES AND SCOPE

As you learned in Lesson 10, "Getting Started with Functions," functions let you divide your programs into small, manageable pieces. All the functions you have used to this point have been quite simple. As your functions perform more meaningful work, there will be times when the functions must use variables to perform their processing. The variables you declare within a function are *local variables*. Their values, and even the fact that the local variables exist, are known only to the function. In other words, if you declare a local variable named *salary* in the *payroll* function, other functions do not have access to *salary's* value. In fact, other functions have no idea that the variable *salary* exists. This lesson examines variable *scope*, or the program locations where your program knows about and can use a variable. By the time you finish this lesson, you will understand the following key concepts:

- Within a function, you declare local variables just as you would within *main*—by specifying the variable's type and name.

- The variable names you use within a function must be unique only to that function.

- A variable's *scope* defines the program locations the program knows about and where the program can access the variable.

- *Global variables*, unlike local variables, are known throughout your entire program and are accessible within all functions.

- The C++ global resolution operator (::) lets you control a variable's scope.

Declaring local variables within a function is very easy. In fact, you have already done so each time you declared variables within *main*.

DECLARING LOCAL VARIABLES

A *local variable* is a variable your program defines within a function. The variable is *local* to the function because only that function knows that the variable exists and can use the variable. You declare local variables at the start of a function, following the function's opening brace, as shown here:

```
void some_function(void)
  {
    int count;
    float result;
  }
```

The following program, *UseBeeps.CPP*, uses the *sound_speaker* function, which sounds the computer's built-in speaker the number of times the parameter *beeps* specifies. Within the *sound_speaker* function, the local variable *counter* keeps track of the number of times the function has sounded the speaker, as shown here:

```
#include <iostream.h>

void sound_beeps(int beeps)
  {
    for (int counter = 1; counter <= beeps; counter++)
      cout << '\a';
  }
```

95

```
void main(void)
 {
   sound_beeps(2);
   sound_beeps(3);
 }
```

As you can see, the *sound_beeps* function declares the variable *counter* immediately after the function's opening brace. Because the function *sound_beeps* defines *counter*, the variable is local to the *sound_beeps* function, meaning only *sound_beeps* has knowledge of and access to the variable.

UNDERSTANDING NAME CONFLICTS

When you declare local variables within a function, it is very likely that the name of a local variable you declare in one function is the same as a variable name you are using in another function. As briefly discussed, local variables are known only to the function that defines their type and name. As such, if two functions use the same local variable name, there is no conflict. C++ treats each variable name as local to the corresponding function.

The following program, *LclName.CPP*, uses the function *add_values* to add two integer values. The function assigns its result to the local variable *value*. In *main*, however, one of the parameters *main* passes to the function is also named *value*. However, because C++ treats both variables as local to their respective functions, the names do not conflict, as shown here:

```
#include <iostream.h>

int add_values(int a, int b)
 {
   int value;

   value = a + b;

   return(value);
 }

void main(void)
 {
   int value = 1001;
   int other_value = 2002;

   cout << value << " + " << other_value << " = " <<
     add_values(value, other_value) << endl;
 }
```

Understanding Local Variables

Local variables are variables you declare within a function. The local variables' names and values are known only to the function that declares the variable. You should declare local variables at the start of your functions, immediately following the functions' first open brace. The names that you assign to local variables must be unique to their corresponding function. When you declare a local variable within a function, you can initialize the variable using the assignment operator.

UNDERSTANDING GLOBAL VARIABLES

As you have learned, a local variable is declared within and known only to a specific function. In addition to local variables, C++ lets your programs declare *global variables*, which are known throughout your entire program (globally to all functions). To declare a global variable, you simply place the variable declaration at the start of your program, outside of any functions, as shown here:

```
int some_global_variable;————————————— Global variable declaration


void main(void)
  {
     // Program statements would be here
  }
```

The following program, *Global. CPP*, uses a global variable named *number*. Each of the functions in the program can use (or change) the global variable's value. In this example, each function displays the variable's current value and then increments the value by 1:

```
#include <iostream.h>

int number = 1001;

void first_change(void)
  {
    cout << "number's value in first_change " << number << endl;
    number++;
  }

void second_change(void)
  {
    cout << "number's value in second_change " << number << endl;
    number++;
  }

void main(void)
  {
    cout << "number's value in main " << number << endl;
    number++;
    first_change();
    second_change();
  }
```

When you compile and execute the *Global. CPP* program, your screen will display the following output:

```
C:\> Global   <ENTER>
number's value in main 1001
number's value in first_change 1002
number's value in second_change 1003
```

As a general rule, your programs should avoid the use of global variables. Because any function can change a global variable's value, it is hard to keep track of all the possible functions within which a global variable's value may change.

Instead, your programs should declare the variable within *main* and then pass (as a parameter) the variable to the functions that need the variable. (Remember, C++ places a temporary copy of the variable's value on the stack, leaving the original unchanged.)

WHEN GLOBAL AND LOCAL VARIABLE NAMES CONFLICT

You should avoid using global variables whenever possible. However, should your program use a global variable, there may be times when the global variable name conflicts with that of a local variable. When such conflicts occur, C++ gives the local variable precedence. In other words, the program assumes each reference to the name in conflict corresponds to the local variable.

There may be times, however, when you must access a global variable whose name conflicts with a local variable. In such cases, when you want to use the global variable your programs can use the C++ *global resolution operator* (::). For example, assume that you have local and global variables named *number*. When your functions want to use the local variable *number*, the function simply refers to the variable, as shown here:

```
number = 1001;    // Local variable reference
```

When the function wants to reference the global variable, on the other hand, the program uses the global resolution operator, as shown here:

```
::number = 2002;    // Global variable reference
```

The following program, *GlobLoca.CPP*, uses the global variable *number*. In addition, the function *show_numbers* uses a local variable named *number*. The function uses the global resolution operator to access the global variable, as shown here:

```
#include <iostream.h>

int number = 1001;  // Global variable

void show_numbers(int number)
  {
    cout << "Local variable number contains " << number << endl;
    cout << "Global variable number contains " << ::number << endl;
  }

void main(void)
  {
    int some_value = 2002;

    show_numbers(some_value);
  }
```

When you compile and execute the *GlobLoca.CPP* program, your screen will display the following output:

```
C:\> GlobLoca   <ENTER>
Local variable number contains 2002
Global variable number contains 1001
```

As you can see, your programs can select the global or local variable using the global resolution operator. However, as you might note, the use of global and local variables can cause confusion, which, in turn, can cause errors. As a result, you should avoid the use of global variables whenever possible.

Understanding Global Variables

A global variable is a variable whose name and value is known throughout your program. To create a global variable, you declare the variable near the top of your source file and outside of any functions. All the functions that follow the variable's declaration can use the global variable. However, because their misuse can lead to errors, you should avoid the use of global variables whenever possible.

UNDERSTANDING A VARIABLE'S SCOPE

As you read books and magazine articles on C++, you may encounter the term *scope*, which defines the locations in a program where a variable's name has meaning (and is therefore usable). For a local variable, the C++ restricts the variable's scope to the function that declares the variable. Global variables, on the other hand, are known throughout your entire program. As a result, global variables have a larger scope.

WHAT YOU MUST KNOW

In this lesson you learned how to declare local variables within your functions. As your functions perform more meaningful work, they will require local variables. This lesson also presented global variables, whose names and values are known throughout your entire program. Because they can lead to errors that are difficult to detect, you should avoid using global variables whenever possible. In Lesson 14, "Overloading Functions," you will learn how C++ lets your programs declare two or more functions with the same name, but which support different parameters or a different return type. By "overloading" a function name in this way, you simplify the function's use. Before you continue with Lesson 14, however, make sure that you have learned the following key concepts:

- ☑ Local variables are variables your program declares within a function.

- ☑ Local variables are known only to the function that declares them.

- ☑ Two or more functions can use the same local variable name with no conflicts.

- ☑ A global variable is a variable whose name and value is known throughout your program.

- ☑ To create a global variable, declare the variable near the top of your source file, outside of any functions.

- ☑ Because any function can readily change a global variable's value, global variables introduce the possibility for errors within your program that can be difficult to detect. Therefore, you should avoid using global variables.

LESSON 14

OVERLOADING FUNCTIONS

When you define functions within your programs, you must specify the function's return type, as well as the number and type of each parameter. In the past, if you programmed in the C programming language and you had a function named *add_values* that worked with two integer values, and you wanted to use a similar function to add three integer values, you had to create a function with a different name. To do so, you might use the functions *add_two_values*, *add_three_values*, and so on. Likewise, if you wanted to use a function to add two values of type *float*, you needed yet another function with its own name.

To eliminate your need to duplicate functions, C++ lets you define multiple functions with the same name. When you compile your program, the C++ compiler examines the number of arguments each function uses and then calls the correct function. The process of providing multiple functions for the compiler to select is *overloading*. This lesson examines how you overload functions within your programs. By the time you finish this lesson, you will understand the following key concepts:

- Function overloading lets you use the same function name with different parameter types or a different number of parameters.

- To overload functions within your program, you simply define two functions of the same name and return type that differ by their number or type of parameters.

Function overloading is a C++ feature that is not available in the C programming language. As you will learn, overloading functions is convenient and can improve your program's readability.

GETTING STARTED WITH FUNCTION OVERLOADING

Function overloading lets your programs define multiple functions with the same name and return type. The following program, *Overload.CPP*, for example, overloads the *add_values* function. The first function definition adds two values of type *int*. The second function definition adds three values. During compilation, the C++ compiler determines the correct function to use:

```
#include <iostream.h>

int add_values(int a, int b)
 {
   return(a + b);
 }

int add_values(int a, int b, int c)
 {
   return(a + b + c);
 }

void main(void)
 {
   cout << "200 + 801 = " << add_values(200, 801) << endl;
   cout << "100 + 201 + 700 = " << add_values(100, 201, 700) << endl;
 }
```

As you can see, the program defines two functions named *add_values*. The first function adds two values of type *int*, while the second adds three values. You do not have to do anything special to warn the compiler about overloading. Instead, you just define and use the functions. The compiler, in turn, will figure out which function definition to use, based on parameters the program supplies. In a similar way, the following program, *Msg_Ovr.CPP*, overloads the function *show_message*. The first function displays a default message if the program passes no parameter to the function. The second function displays the message the program passes to the function, and the third function displays two messages:

```cpp
#include <iostream.h>

void show_message(void)
  {
    cout << "Default message: Rescued by C++" << endl;
  }

void show_message(char *message)
  {
    cout << message << endl;
  }

void show_message(char *first, char *second)
  {
    cout << first << endl;
    cout << second << endl;
  }

void main(void)
  {
    show_message();
    show_message("I've Been Rescued!");
    show_message("C++ is not so hard!", " Overloading is cool!");
  }
```

WHEN TO USE OVERLOADING

One of the most common uses of overloading is to use a function to obtain a result, even though parameter types might differ. For example, assume your program has a function named *day_of_week*, which returns the current day of week (0 for Sunday, 1 for Monday, and so on, up to 6 for Saturday). Your program might overload the function so that it returns the correct day of the week if the program passes it a Julian day as a parameter or if the program passes the function a value for the current day, month, and year, as shown here:

```cpp
int day_of_week(int julian_day)
  {
    // Statements
  }

int day_of_week(int month, int day, int year)
  {
    // Statements
  }
```

As you examine the C++ object-oriented capabilities later lessons present, you will use function overloading to increase your program's capabilities.

Function Overloading Improves Your Program's Readability

Function overloading in C++ lets your programs define multiple functions which use the same name. The overloaded functions must differ by the number and type of their parameters. Prior to C++ support for function overloading, C programs had to create multiple, similarly named functions within their programs. Unfortunately, programmers who must use the functions must remember which function corresponded to which parameter combination. Function overloading, on the other hand, simplifies the programmer's task by requiring the programmers to remember only one function name.

WHAT YOU MUST KNOW

Function overloading lets your programs specify multiple definitions for the same function. When you compile a program, the C++ compiler will determine which function to use, based on the number and type of parameters the program passes to the function. In this lesson, you learned that you can easily overload functions within your programs. In Lesson 15, "Using C++ References," you will learn how C++ references simplify the process of changing parameters within functions. Before you continue with Lesson 15, however, make sure that you have learned the following key concepts:

☑ Function overloading lets your programs provide multiple "views" to the same function within your program.

☑ To overload a function within your program, you simply define two or more functions with the same name that differ by the number or type parameters they support.

☑ During compilation, the C++ compiler will determine which overloaded function to call, based on the number and type of parameters the program passes to the function.

☑ Function overloading simplifies programming by letting programmers work with only one function name when their programs must accomplish a specific task.

LESSON 15

USING C++ REFERENCES

In Lesson 11, "Changing Parameter Values," you learned how to change parameters within functions using pointers. As you learned, to use pointers you have to precede pointer variable names with an asterisk. The use of pointers within C++ programs is a carryover from the C programming language. To simplify changing parameters within functions, C++ provides *references*. As you will learn in this lesson, a reference is an alias (or second name) your programs can use to reference a variable. Using a reference to a variable within a function, you can change a parameter's value without having to use pointers and pointer notation. By the time you finish this lesson, you will understand the following key concepts:

- To declare and initialize a reference within your program, declare a variable, placing an ampersand (&) immediately after the variable's type, and then use the assignment operator to assign the alias, such as *int& alias_name = variable;*.

- Your programs can pass a reference to a function as a parameter, which, in turn, can change the corresponding parameter's value without the use of pointers.

- Within a function, you must declare the parameter as a reference by placing an ampersand (&) after the parameter's type—you can then change the parameter's value within the function without using pointers.

In functions that use pointers, novice C++ programmers are often confused about when they must use the asterisk (*) to "dereference" a pointer and when they should precede a variable name with the address operator (&). As you will learn, using references makes changing parameter values within a function very easy.

A REFERENCE IS AN ALIAS

A C++ *reference* lets your programs create an alias (or second name) for a variable in your program. To declare a reference within your program, you specify a type immediately followed by an ampersand (&) character. When you declare a reference, you must immediately assign the variable the reference will alias, as shown here:

```
int& alias_name = variable;                              Reference declaration
```

After you declare a reference, your program can use the variable or the reference, interchangeably, as shown here:

```
alias_name = 1001;

variable = 1001;
```

The following program, *Show_Ref.CPP*, creates a reference named *alias_name* and assigns the alias the variable *number*. The program then uses both the reference and the variable, as shown here:

```
#include <iostream.h>

void main(void)
  {
    int number = 501;
    int& alias_name = number;       // Create the reference
```

```
   cout << "The variable number contains " << number << endl;
   cout << "The alias to number contains " << alias_name << endl;

   alias_name = alias_name + 500;

   cout << "The variable number contains " << number << endl;
   cout << "The alias to number contains " << alias_name << endl;
 }
```

As you can see, the program adds the value 500 to the reference *alias_name*. Therefore, the program also adds 500 to the corresponding variable *number* for which the reference serves as an alias or second name. When you compile and execute the *Show_Ref.CPP* program, your screen will display the following output:

```
C:\> Show_Ref   <ENTER>
The variable number contains 501
The alias to number contains 501
The variable number contains 1001
The alias to number contains 1001
```

As a general rule, using a reference in the way just shown makes programs difficult for programmers to understand. However, as you will see, using a reference also makes its very easy for your programs to change a parameter within a function.

Declaring a Reference Within Your Programs

 A C++ reference is an alias (or second name) your programs can use to refer to a variable. To declare a reference, you place the ampersand (&) character immediately after the variable's type and then specify the reference name, followed by an equal sign and the name of the variable the reference is to alias, as shown here:

```
float& salary_alias = salary;
```

USING REFERENCES AS PARAMETERS

The fundamental purpose of a reference is to simplify the process of changing parameters within a function. The following program, *Referenc.CPP*, assigns a reference named *number_alias* to the variable *number*. The program passes the reference variable to the *change_value* function, which assigns the variable the value 1001, as shown here:

```
#include <iostream.h>

void change_value(int &alias)
  {
    alias = 1001;
  }

void main(void)
  {
    int number;
    int& number_alias = number;
```

```
    change_value(number_alias);

    cout << "The variable number contains " << number << endl;
}
```

As you can see, the program passes the reference to the *change_value* function. If you examine the function declaration, you will find that *change_value* declares the parameter *alias* as a reference to a value of type *int*:

```
void change_value(int& alias)
```

Within the *change_value* function the code can change the parameter's value without having to use a pointer. As a result, the function code does not use an asterisk (*), and operations within the function become easier for you to understand.

Use Comments to Explain References Within Your Programs

Most C++ programmers are familiar with the C programming language, and they are familiar with using pointers within functions that must change a parameter's value. As a result, when these programmers do not see pointers within functions that use references, they might assume the parameter values do not change. To prevent such misunderstandings, make sure you place several comments before and within functions that change parameters using references. When you do so, C programmers will better understand your function's processing.

LOOKING AT A SECOND EXAMPLE

In Lesson 11, you used the following function to swap two floating-point values:

```
void swap_values(float *a, float *b)
 {
   float temp;

   temp = *a;
   *a = *b;
   *b = temp;
 }
```

As you can see, the function combines pointer variables with non-pointer variables. The following program, *Swap_Ref.CPP*, uses references to the floating-point values to simplify the function:

```
 #include <iostream.h>

void swap_values(float& a, float& b)
  {
   float temp;

   temp = a;
   a = b;
   b = temp;
  }
```

```
void main(void)
 {
    float big = 10000.0;
    float small = 0.00001;
    float& big_alias = big;
    float& small_alias = small;

    swap_values(big_alias, small_alias);
    cout << "Big contains " << big << endl;
    cout << "Small contains " << small << endl;
 }
```

As you can see, when the program uses references to floating-pint values, the *swap_values* function becomes easier to understand. However, your program now has two extra names (the references *big_alias* and *small_alias*) that you have to track.

RULES FOR WORKING WITH REFERENCES

A reference is not a variable. After you assign a value to a reference, the reference cannot change. In addition, unlike pointers, you cannot perform the following operations on references:

- You cannot get the reference's address using the C++ address operator.

- You cannot assign a pointer to a reference.

- You cannot compare reference values using the C++ relational operators.

- You cannot perform arithmetic operations on a reference, such as adding an offset value.

- You cannot change a reference.

As your programs use the C++ object-oriented capabilities, you will revisit references.

Using References to Change Function Parameters

In Lesson 11 you learned that your programs can change the value of parameters within functions by using pointers. Before a function can change a parameter, you must pass the parameter's address to the function. To get the parameter's address, you use the C++ address operator (&). The function, in turn, uses pointer variables (which store a memory address). To declare a pointer variable within the function, you precede the parameter name with an asterisk (*). To change or use the parameter's value within the function, you precede each reference to the parameter name with the C++ indirection operator (*). Unfortunately, many operations within functions combine pointer and non-pointer variables.

C++ references simplify the process of changing function parameters by eliminating statements that combine pointer and non-pointer variables. The processing cost that references add to your program are additional variables that a programmer who is reading your code must recognize and understand.

WHAT YOU MUST KNOW

In this chapter you learned how to use C++ references to create an alias (or second name) for a variable. Using references can simplify functions that change the value of parameters. In Lesson 16, "Providing Default Parameter Values," you will learn that C++ lets you provide default values for function parameters. Should a program omit one or more parameter values within a function call, the function will use the defaults. Before you continue with Lesson 16, however, make sure that you have learned the following key concepts:

☑ A C++ reference is an alias (or second name) for a variable.

☑ To declare a reference, you place the ampersand (&) character immediately after the variable's type, and then specify the reference name followed by an equal sign and the name of the variable the reference is to alias.

☑ After you assign a value to a reference, you cannot change the reference value.

☑ You should include several comments before and within functions that use references to change a parameter's value to ensure that another programmer who is reading your code can recognize the change.

☑ Overusing references can lead to program code that is very difficult for you and other programmers to understand.

LESSON 16

PROVIDING DEFAULT PARAMETER VALUES

As you have learned, C++ lets your programs pass information to functions using parameters. In Lesson 14, "Overloading Functions," you learned that C++ lets you overload functions by providing definitions that support a different number of parameters or even different parameter types. In addition, C++ lets your programs omit parameters when they invoke a function. In such cases, C++ will use default values for the missing parameters. This lesson examines how you define default parameter values within your functions. By the time you finish this lesson, you will understand the following key concepts:

- C++ lets your programs specify default parameter values.

- You specify default parameter values within the function header when you define the function.

- If a function call omits one or more parameter values, C++ will use the default values.

- When a function call omits a parameter, the call must omit all the parameters that follow.

Providing default values for parameters simplifies a program's use of your functions. As a result, more programs may use the function—which increases the function's *reuse.*

SPECIFYING DEFAULT VALUES

Providing default values for function parameters is very easy. In short, you use the C++ assignment operator to assign a value to a parameter that appears in the function definition, as shown here:

```
void some_function(int size=12, float cost=19.95)
{
   // Function statements
}
```

The following program, *Defaults.CPP,* assigns default values to the parameters *a, b,* and *c* within the *show_parameters* function. The program then invokes the function, first specifying no parameter values, then a value for *a,* values for *a* and *b,* and, finally, values for all three parameters:

```
#include <iostream.h>

void show_parameters(int a=1, int b=2, int c=3)
{
   cout << "a " << a << " b " << b << " c " << c << endl;
}

void main(void)
{
   show_parameters();
   show_parameters(1001);
   show_parameters(1001, 2002);
   show_parameters(1001, 2002, 3003);
}
```

When you compile and execute the *Defaults.CPP* program, your screen will display the following output:

```
C:\> Defaults   <ENTER>
a 1 b 2 c 3
a 1001 b 2 c 3
a 1001 b 2002 c 3
a 1001 b 2002 c 3003
```

As you can see, the function uses the default parameter values as needed.

RULES FOR OMITTING PARAMETER VALUES

When your program omits parameters for a function that provides default values, your program must omit all the parameters that follow. In other words, you cannot omit a middle parameter. In the case of the previous program, if your program wants to omit the value for parameter *b* in *show_parameters*, your program must omit parameter *c's* value as well. You cannot specify a value for *a* and *c*, omitting *b's* value.

Specifying Default Parameter Values

When you define a function, C++ lets you specify default values for one or more parameters. Should the program omit one or more parameters when it later calls the function, the function will use its default values. To assign a default value to a parameter, simply use the assignment operator within the function definition. For example, the following *payroll* function specifies default values for the *hours* and *rate* parameters:

```
float payroll(int employ_id, float hours = 40, float rate = 5.50)
  {
     // statements
  }
```

When a program omits one parameter within a function call, the program must omit the values for all the parameters that follow.

WHAT YOU MUST KNOW

In this lesson you learned that C++ lets you specify default values for function parameters. Should a program later omit one or more parameters within a function call, the function will use the default values. In later lessons, when your programs start using C++ object-oriented features, you will use default parameters to initialize different variables within a class. As you have learned, a variable lets you store a value of a specific type (*int*, *float*, and so on). In Lesson 17, "Using Constants and Macros," you will learn how constants and macros can simplify your programming or produce more readable code. Before you continue with Lesson 17, make sure that you have learned the following key concepts:

☑ To assign default values to function parameters, you use the C++ assignment operator to assign a value to the parameter within the function definition.

☑ When a program omits parameter values during a function call, the function will use the default values.

☑ When a program omits a parameter value, the program must omit the values for all parameters that follow—programs cannot omit a middle parameter's value.

☑ By specifying default parameter values, you may make your functions easier to use and possibly better suited for another programmer's use.

109

LESSON 17

USING CONSTANTS AND MACROS

As your programs increase in complexity, they may also become more difficult for other programmers to understand. To improve your program's readability, C++ supports named constants and macros. Using named constants you can replace a numeric value, such as 50, within your source code, with a meaningful constant, such as *CLASS_SIZE*. When another programmer reads your code, he or she will not have to guess what the numeric value 50 means. Instead, each time the programmer sees *CLASS_SIZE*, he or she will know it contains a value that corresponds to the number of students in a class. Likewise, using a *macro*, your programs can replace complex equations such as the following with a meaningful macro name:

```
result = (x*y-3) * (x*y-3) * (x*y-3);
```

Using a macro, your programs can replace this complex equation with a function-call like macro named *CUBE*, as shown here:

```
result = CUBE(x*y-3);
```

In this example, the macro not only improves the readability of your code, it also simplifies your statement, reducing the chance of error. This lesson examines named constants and macros in detail. By the time you finish this lesson, you will understand the following key concepts:

- To make your programs easier to read, replace numeric values with more meaningful named constants.

- Using named constants throughout your program in place of numeric values can make your program easier to modify in the future.

- C++ lets your programs replace equations with meaningful macro names.

- Before the C++ compiler compiles your program, it uses a special program, called the *preprocessor*, to replace each constant or macro with its corresponding value.

- Macros execute faster than functions, but they also increase the size of your executable program.

- Most C++ compilers predefine constants or macros you can use throughout your program.

USING NAMED CONSTANTS

A *named constant* is simply a name to which you assign a constant value, which, unlike a variable's value, cannot change as your program executes. You create named constants using the *#define* preprocessor directive (a directive is a special instruction for the compiler's preprocessor). For example, the following statement defines the *CLASS_SIZE* named constant as the value 50:

```
#define CLASS_SIZE 50
```

To differentiate named constants from variables, most programmers will use uppercase letters for named constants. The following program, *Constant.CPP*, for example, defines and displays the *CLASS_SIZE* named constant:

```
#include <iostream.h>

#define CLASS_SIZE 50   // Number of students in the class

void main(void)
 {
   cout << "CLASS_SIZE constant is " << CLASS_SIZE << endl;
 }
```

As you can see, the program defines the constant using the *#define* directive near the top of the source code. After you define a constant, you can use the constant's value throughout your program by simply referencing the constant value's name.

Note: The previous constant definitions do not end with a semicolon. If you include the semicolon at the end of a definition, the preprocessor will include the semicolon within your definition. For example, if you placed a semicolon after the value 50 in the previous program's #define directive, the preprocessor would later substitute the value 50 and the semicolon (50;) for each occurrence of the constant CLASS_SIZE, which would very likely generate a syntax error.

Understanding Preprocessor Directives

Before the C++ compiler compiles your program, the compiler runs a special program named the preprocessor. The preprocessor examines your program for lines that begin with a #, such as *#include* or *#define*. If, for example, the preprocessor encounters a *#include* directive, the preprocessor includes the specified file within your source file, just as if you had typed in the file's contents when you created your source code. Every program you have created in this book has used the *#include* directive to instruct the preprocessor to include the contents of the header file *iostream.h* in your source file. If the preprocessor encounters a *#define* directive, the preprocessor creates a named constant or macro. Later, when the preprocessor encounters a constant or macro name, the preprocessor replaces the name with the value your program specifies using the *#define* directive.

When you define constants within your programs, C++ does not restrict your constants to numeric values. Rather, you can use constants to store strings and floating-point numbers as well. For example, the following program, *BookInfo.CPP*, uses the *#define* directive to create three constants that contain information about this book:

```
#include <iostream.h>

#define TITLE "Rescued by C++, Third Edition"
#define LESSON 17
#define PRICE 29.95

void main(void)
 {
   cout << "Book Title: " << TITLE << endl;
   cout << "Current Lesson: " << LESSON << endl;
   cout << "Price: $" << PRICE << endl;
 }
```

When you compile and execute the *BookInfo.CPP* program, your screen will display the following output:

```
C:\> BookInfo   <ENTER>
Book Title: Rescued by C++, Third Edition
Current Lesson: 17
Price: $29.95
```

Using #define to Create Named Constants

To improve your program's readability, you should replace numeric values that appear throughout your program with meaningful named constants. To define a named constant, your programs use the *#define* preprocessor directive. Your should place your constants near the top of your source file. Also, to differentiate constants from variables, most programmers use uppercase letters for constant names. For example, the following *#define* directive creates a constant named *SECONDS_PER_HOUR*:

```
#define SECONDS_PER_HOUR 3600
```

When you compile your program, the C++ preprocessor will replace each occurrence of the name *SECONDS_PER_HOUR* with the numeric value 3600. Note that the constant definition does not end with a semicolon. If you include a semicolon following the 3600, the C++ preprocessor would later substitute each occurrence of the *SECONDS_PER_HOUR* with the value and semicolon (3600;), very likely causing a syntax error.

USING NAMED CONSTANTS TO SIMPLIFY CODE CHANGES

In addition to making your programs easier to read, named constants make your programs easier to modify. For example, the following code fragment makes several references to the number 50 (the number of students in a class):

```cpp
#include <iostream.h>

void main(void)
  {
    int student;

    for (student = 0; student < 50; student++)
      get_test_score(student);

    for (student = 0; student < 50; student++)
      calculate_grade(student);

    for (student = 0; student < 50; student++)
      print_grade(student);
  }
```

Assume, for example, that the number of students in the class grows to 55. In such a case, you must edit the previous program and replace each occurrence of the value 50 with the value 55. As an alternative, the following program uses the named constant *CLASS_SIZE*:

```
#include <iostream.h>

#define CLASS_SIZE 50

void main(void)
  {
    int student;

    for (student = 0; student < CLASS_SIZE; student++)
      get_test_score(student);

    for (student = 0; student < CLASS_SIZE; student++)
      calculate_grade(student);

    for (student = 0; student < CLASS_SIZE; student++)
      print_grade(student);
  }
```

In this example, to change the class size throughout the program, you must change only the line that contains the *#define* directive that defines the constant:

```
#define CLASS_SIZE 55
```

REPLACING EQUATIONS WITH MACROS

When your programs perform real-world operations, it is common for your code to include complex expressions, as shown here:

```
result = (x*y-3) * (x*y-3) * (x*y-3);
```

In this example, the program calculates the cube of the expression (x*y-3). To make your program easier to read and to reduce the chance of introducing an error by mistyping the expression, you can create a macro named *CUBE* that your program can use as shown here:

```
result = CUBE(x*y-3);
```

Again, it is common practice for programmers to use uppercase letters for macro names to distinguish macros from functions.

To create a macro, you use *#define* preprocessor directive. The following statement, for example, creates the macro *CUBE*:

```
#define CUBE(x) ((x)*(x)*(x))
```

As you can see, the program uses the *#define* statement to create the *CUBE* macro to multiply the parameter *x* times itself, twice. The following program, *ShowCube.CPP*, uses the *CUBE* macro to display the cube of the values 1 through 10:

```
#include <iostream.h>

#define CUBE(x) ((x)*(x)*(x))
```

113

```
void main(void)
 {
    for (int i = 1; i <= 10; i++)
      cout << i << " cubed is " << CUBE(i) << endl;
 }
```

When you compile the *ShowCube.CPP* program, the C++ preprocessor replaces each occurrence of the macro *CUBE* with the corresponding macro definition. In other words, the preprocessor's macro substitution results in the following code:

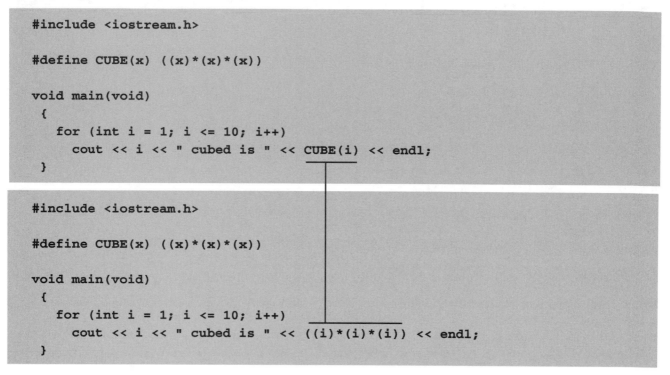

```
#include <iostream.h>

#define CUBE(x) ((x)*(x)*(x))

void main(void)
 {
    for (int i = 1; i <= 10; i++)
      cout << i << " cubed is " << CUBE(i) << endl;
 }
```

```
#include <iostream.h>

#define CUBE(x) ((x)*(x)*(x))

void main(void)
 {
    for (int i = 1; i <= 10; i++)
      cout << i << " cubed is " << ((i)*(i)*(i)) << endl;
 }
```

Note that the previous macro definition placed the parameter *x* within parentheses using *((x)*(x)*(x))* as opposed to *(x*x*x)*. When you create macros, you should place parameters within parentheses in this way to ensure that C++ evaluates your expressions the way you want. As you may recall from Lesson 6, "Performing Simple Operations," C++ uses an operator precedence to determine the order in which it performs arithmetic operations. Assume, for example, a program uses the *CUBE* macro with the equation *3+5-2,* as shown here:

```
result = CUBE(3+5-2);
```

If the macro groups the parameter within parentheses, the preprocessor will generate the following statement:

```
result = ((3+5-2) * (3+5-2) * (3+5-2));
```

However, if the macro definition omits the parentheses, the preprocessor will generate the following statement:

```
result = (3+5-2*3+5-2*3+5-2);
```

If you take time to calculate each equation, you will find that the results differ. By grouping macro arguments within parentheses, you avoid such errors.

HOW MACROS DIFFER FROM FUNCTIONS

A macro definition is not a function. When your program uses a function, only one copy of the function's statements resides in your executable program. Each time your program calls the function, your program pushes parameters onto the stack and then jumps to the function's code. When the function completes, your program pops values from the stack and jumps back to the statement that immediately follows the function call.

In the case of a macro, on the other hand, the preprocessor replaces each macro reference in your code with the corresponding macro definition. For example, if the previous program used the *CUBE* macro in 100 different locations, the preprocessor would substitute in the macro code 100 different times. With a macro, you don't have the overhead of a function call (the overhead to push and pop parameters on and off the stack and the overhead to jump to and return from the function code). That's because, with a macro, the preprocessor places the corresponding statements inline. However, because the preprocessor replaces each macro reference with the corresponding code, macros increase the size of your executable program.

MACRO USE IS VERY FLEXIBLE

You can use macros within your programs in a wide variety of ways. However, keep in mind that your goal when using macros is to simplify coding and to improve your program's readability. The following program, *MacDelay.CPP*, illustrates macro flexibility. In addition, the program may give you a better picture of how the preprocessor replaces a macro name with its corresponding statements:

```
#include <iostream.h>

#define delay(x) { \
                cout << "Delaying for " << x << endl; \
                for(long int i=0; i < x; i++) \
                  ; \
              }

void main(void)
 {
   delay(100000L);

   delay(200000L);

   delay(300000L);
 }
```

In this case, because the macro definition spans several lines, the definition places a single back slash (\) at the end of each line that continues to the next. When the preprocessor encounters a macro reference, the preprocessor will replace the reference with each statement that appears in the macro definition.

WHAT YOU MUST KNOW

Macros and named constants exist to improve your program's readability and to simplify your programming. This lesson examined how you create and use named constants and macros within your code. In Lesson 18, "Storing Multiple Values in Arrays," you will learn how to store multiple values of the same type within an *array*. For example, your program might store 100 student test scores or 50 stock prices. Using arrays, storing and using such values is very easy. Before you continue with Lesson 18, however, make sure you have learned the following key concepts:

☑ Macros and named constants make your programs easier to read by replacing complex equations and numeric constants with meaningful names.

☑ By replacing numeric values with named constants throughout your program, you reduce the number of changes you must make to your program at a later time, should the constant's value change.

☑ During compilation, the C++ compiler uses a special program called the preprocessor to replace each constant or macro with its corresponding value.

☑ Macros execute faster than functions, but also increase the size of your executable program.

☑ If a macro's definition exceeds one line, you can place one back slash character (\) at the end of the line to inform the preprocessor that the definition continues on the next line.

ESSENTIAL C/C++ LINKS

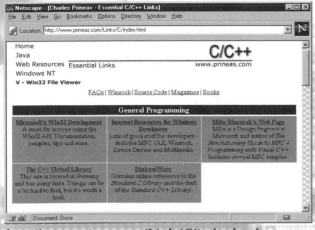

http://www.prineas.com/Links/C/index.html

FREQUENTLY ASKED QUESTIONS

Netscape - [C++ FAQ LITE, Marshall Cline]

File Edit View Go Bookmarks Options Directory Window Help

Netsite: http://www.cerfnet.com/~mpcline/On-Line-C++-FAQs/

C++ FAQ LITE — Frequently Asked Questions
(Copyright © 1991-96, Marshall Cline, cline@parashift.com)

Improve network response:

- Use the closest mirror site: USA, U.K., Germany, Finland, Taiwan.
- Browse the FAQ off-line: Use the one-click download feature to get your own local copy.
  NEW

Look up stuff:

- Look up a generic topic with the table of contents (35 coarse-grained topics).
- Look up a specific topic with the extensive subject index (707 fine-grained topics).
- Look up the changed topics with the chain of recent changes (29 changes for 1/97 and

Document: Done

http://www.cerfnet.com/~mpcline/On-Line-C++FAQs/

KNOW-HOW FOR DOS

Netscape - [Order Information]

File Edit View Go Bookmarks Options Directory Window Help

Location: http://home.istar.ca/~stepanv/kh_dos.shtml

LIST OF PRODUCTS

KNOW-HOW for DOS

Software development kit

Library in source codes (Borland C++ 3.x)

KNOW-HOW for DOS / Windows

is a Library (all source codes included, Borland C++ 3.x). It was designed mostly for DOS, but most of it will work under the Win 3.1 as well. No testing was performed for Win95, however there should be no major problems.

The library is supplied as the set of .H and .CPP files, with the project files for Borland C++. Every CPP file contains the remarked function main(), that illustrates the use of this file's features. Code is heavily commented, tutorials in the form of text files are included.

Document: Done

http://home.istar.ca/~stepanv/kh_dos.shtml

OBJECT-ORIENTED PROGRAMMING

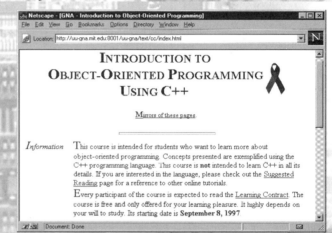

http://uu-gna.mit.edu:8001/uu-gna/text/cc/index.html

C++ ANNOTATIONS VERSION 4.2.1

Netscape - [C++ Annotations Version 4.2.1]

File Edit View Go Bookmarks Options Directory Window Help

Location: http://www.icce.rug.nl/docs/cplusplus/cplusplus.html

C++ Annotations Version 4.2.1

Frank B. Brokken and Karel Kubat

**ICCE, University of Groningen
Westerhaven 16, 9718 AW Groningen
Netherlands
Published at the University of Groningen
ISBN 90 367 0470 7**

1994 - 1997

This document is intended for knowledgeable users of C who would like to make the transition to C++. It is a guide for Frank's C/C++ programming courses, which are given yearly at the University of Groningen. As such, this document is not a complete C/C++ handbook. Rather, it serves as an addition

Document: Done

http://www.icce.rug.nl/docs/cplusplus.html

1997 C++ PUBLIC REVIEW

Netscape - [1997 C++ Public Review Document]

File Edit View Go Bookmarks Options Directory Window Help

Location: http://www.maths.warwick.ac.uk/cpp/pub/wp/html/cd2/index.html

1997 C++ Public Review Document	0 Accredited Standards Committee* X3, INFORMATION PROCESSING SYSTEMS

Search

Introduction

Coversheet
Foreword
Table of Contents
1: General

Core Language

2: Lexical conventions
3: Basic concepts
4: Standard conversions

0

Working Paper for
Draft Proposed International Standard for Inf
Programming Language C++

Abstract

0 This document specifies the form and establishes
programs expressed in the programming language C

Document: Done

http://www.maths.warwick.ac.uk/cpp/pub/wp/html/cd2/index.html

VISUAL C++ DEVELOPERS JOURNAL

http://www.vcdj.com/default.htm

LEARN C/C++ TODAY

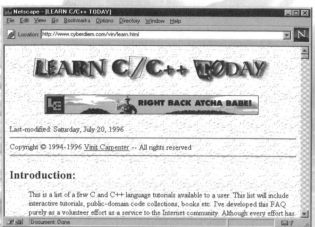

http://www.cyberdiem.com/vin/learn.html

C/C++ RESOURCES

http://www.geocities.com/SiliconValley/Pines/1060/—cpp.html

THE STANDARD TEMPLATE LIBRARY

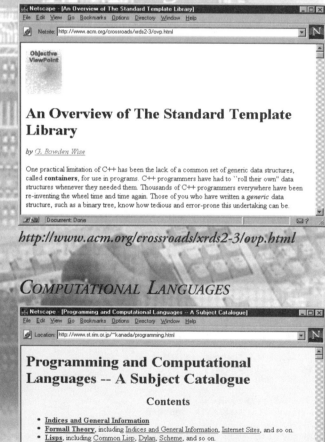

http://www.acm.org/crossroads/xrds2-3/ovp.html

PO - C/C++

http://www.ulu.fi/~sisasa/oasis/oasis-cc++.html

COMPUTATIONAL LANGUAGES

http://www.st.rim.or.jp/~kanada/programming.html

SECTION THREE

STORING INFORMATION USING ARRAYS AND STRUCTURES

As you have learned, a variable can store a value of a specific type. In the C++ programs you created in this book's first two sections, variables have stored only one value. As your programs become more complex, there will be times when your programs must work with many values at the same time. For example, your program might work with 100 test scores, 30 stock prices, or the names and addresses of your company's 5,000 employees. In this section you will learn how to use different C++ data types to store multiple values within the same variable. As you will learn, using one variable to store multiple values is very convenient. The lessons in this section include:

LESSON 18

STORING MULTIPLE VALUES IN ARRAYS

As your programs execute, they store information in variables. Until now, the variables within your programs have only stored one value at a time. In many cases, however, your programs must store multiple values, such as 50 test scores, 100 book titles, or 1,000 filenames. To store multiple values, your programs use a special data structure called an *array*. In short, an array is simply a variable that can hold multiple values of the same type, such as 10 values of type *int*. To declare an array, your programs specify the array type, name, and number of items the array will store. This lesson examines how your programs declare and later store and retrieve information within an array. By the time you finish this lesson, you will understand the following key concepts:

- An *array* is a data structure that lets a single variable store multiple values of the same type.

- When you declare an array, you must specify the type of value the array will store as well as the number of items the array will hold (called *array elements*).

- Each element within an array must be the same type, such as *int*, *float*, or *char*.

- To store a value within an array, you specify the element number within the array at which you want to store the value. For example, the array's first element is element 0, the second is element 1, and so on.

- To access a value stored within an array, your programs specify the array name and the element number, placing the element number within left-and-right brackets, such as *scores[3]*.

- When your program declares an array, it can use the assignment operator to initialize array elements.

- Your programs can pass array variables to functions just as they would any parameter.

C++ programs make extensive use of arrays. In Lesson 19, "Understanding Character Strings," you will work with character strings (such as book title, filename, and so on) and arrays of characters.

DECLARING AN ARRAY VARIABLE

An *array* is a variable that can store one or more values of the same type. Like the variables your programs have used to this point, an array must have a type (such as *int*, *char*, or *float*) and a unique name. In addition, you must specify the number of values the array will store. All the values you store in an array must be the same type. In other words, your program cannot place values of type *float*, *char*, and *long* in the same array. The following declaration creates an array named *test_scores*, that can hold 100 integer test scores:

```
int test_scores[100];        Array type
                             Array size
```

When the C++ compiler encounters the array declaration, the compiler will allocate enough memory for the variable to hold 100 values of type *int*. The values your programs store in an array are *array elements*.

Arrays Store Multiple Values of the Same Type

As your programs become more complex, they will work with multiple values of the same type. For example, your program might store the prices for 50 parts, the ages of 100 employees, or 25 stock prices. Rather than forcing your program to work with 50, 100, or 25 uniquely named variables, C++ lets your programs define a single variable, an array, that can store multiple, related values.

To declare an array, you must specify the array type, a unique name, and the number of elements the array will hold. For example, the following statements declare three different arrays:

```
float part_cost[50];
int employee_age[100];
float stock_prices[25];
```

ACCESSING ARRAY ELEMENTS

As you have learned, an array lets your programs store multiple values within the same variable. To access specific values your program stores in the array, you use an *index value* that points to the desired element. For example, to access the first element in the array *test_scores*, you would use the index value 0 (*test_scores[0]*). To access the second element, you would use the index value 1 (*test_scores[1]*). Likewise, to access the third value, you would use an index value of 2 (*test_scores[2]*). As shown in Figure 18.1, C++ programs always index the first array element with 0 and the last array element with a value one less than the size of the array.

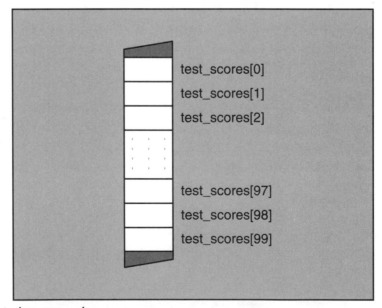

Figure 18.1 How C++ indexes array elements.

It is important for you to remember that C++ always uses 0 to index the first array element and the array size minus 1 for the last element. The following program, *Array.CPP*, creates an array named *values*, that holds five integer values. The program then assigns the elements the values 100, 200, 300, 400, and 500:

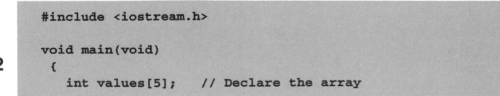

```
#include <iostream.h>

void main(void)
{
    int values[5];    // Declare the array
```

```
   values[0] = 100;
   values[1] = 200;
   values[2] = 300;
   values[3] = 400;
   values[4] = 500;

   cout << "The array contains the following values" << endl;
   cout << values[0] << ' ' << values[1] << ' ' << values[2] << ' ' <<
     values[3] << ' ' << values[4] << endl;
}
```

As you can see, the program assigns the first value to element 0 (*values[0]*). Also, the program assigns the last value to element 4 (the array size (5) minus 1).

Use an Index Value to Access Array Elements

 An array lets your programs store multiple values of the same type within the same variable. To access specific values within the array, your programs use an index value. In short, the index value specifies the array element you desire. All C++ arrays start with the element 0. The following statement, for example, assigns the value 100 to the first element of the array *scores*:

```
   scores[0] = 100;
```

When your program declares an array, your program specifies the number of elements the array can store. For example, the following statement declares an array capable of storing 100 values of type *int*:

```
   int scores[100];
```

In this example, the array elements are *scores[0]* through *scores[99]*.

USING AN INDEX VARIABLE

When your programs use arrays, a common operation is to use an index variable to access the array elements. For example, assuming the index variable *i* contains the value 3, the following statement assigns the value 400 to *values[3]*:

```
   values[i] = 400;
```

The following program, *ShowArra.CPP*, uses the index variable *i* within a *for* loop to display the array elements. The *for* loop initializes *i* to 0, so that it can reference the element *values[0]*, and it ends when *i* is greater than 4 (the array's last element):

```
   #include <iostream.h>

   void main(void)
    {
      int values[5];    // Declare the array
      int i;
```

```
    values[0] = 100;
    values[1] = 200;
    values[2] = 300;
    values[3] = 400;
    values[4] = 500;

    cout << "The array contains the following values" << endl;

    for (i = 0; i < 5; i++)
      cout << values[i] << ' ';
}
```

Each time the *for* loop increments the variable *i*, the program can access the next array element. Experiment with the *ShowArra.CPP* program, changing the *for* loop to the following:

```
    for (i = 4; i >= 0; i—)
      cout << values[i] << ' ';
```

In this case, the program will display the array elements from highest to lowest.

INITIALIZING AN ARRAY AT DECLARATION

As you have learned, C++ lets your programs initialize variables at their declaration. The same is true for arrays. When you declare an array, you can specify initial values by placing the values within left-and-right braces following an equal sign. For example, the following statement initializes the array *values*:

```
  int values[5] = { 100, 200, 300, 400, 500 };
```

In a similar way, the following declaration initializes a floating-point array:

```
  float salaries[3] = { 25000.00, 35000.00, 50000.00 };
```

If you do not specify initial values for each array element, most C++ compilers will initialize the element to 0. For example, the following declaration initializes the first three elements of a five-element array:

```
  int values[5] = { 100, 200, 300 };
```

In this example, the statement did not initialize the elements *values[3]* and *values[4]*. Depending on your compiler, the compiler may assign the elements the value 0. If you do not specify the size of an array you initialize at declaration, C++ will allocate enough memory to hold the number of elements for values you specify. For example, the following declaration creates an array capable of storing four integer values:

```
  int numbers[] = { 1, 2, 3, 4 };
```

PASSING ARRAYS TO FUNCTIONS

Like most variables, your programs will pass arrays to functions. The function might initialize the array, add up the array values, or display the array entries on your screen. When you pass an array to a function, you must specify the array type. You do not need to specify the array size. Instead, you will pass a parameter, such as the variable *number_of_elements*, that specifies the number of elements in the array, as shown here:

```
void some_function(int array[], int number_of_elements);
```

The following program, *ArrayFun.CPP*, passes arrays to the function *show_array*, which uses a *for* loop to display the array values:

```
#include <iostream.h>

void show_array(int array[], int number_of_elements)
  {
    int i;

    for (i = 0; i < number_of_elements; i++)
      cout << array[i] << ' ';

    cout << endl;
  }

void main(void)
  {
    int little_numbers[5] = { 1, 2, 3, 4, 5 };
    int big_numbers[3] = { 1000, 2000, 3000 };

    show_array(little_numbers, 5);
    show_array(big_numbers, 3);
  }
```

As you can see, the program passes the array to the function by name and also specifies a parameter that tells the function the number of elements the array contains, as shown here:

```
show_array(little_numbers, 5);
```

The following program, *GetArray.CPP*, uses the *get_values* function to assign three values to the array *numbers*:

```
#include <iostream.h>

void get_values(int array[], int number_of_elements)
  {
    int i;

    for (i = 0; i < number_of_elements; i++)
      {
        cout << "Enter value " << i << ": ";
        cin >> array[i];
      }
  }

void main(void)
  {
    int numbers[3];

    get_values(numbers, 3);
```

125

```
      cout << "The array values are as follows" << endl;

   for (int i = 0; i < 3; i++)
     cout << numbers[i] << endl;
 }
```

As you can see, the program passes the array to the function by name. The function, in turn, assigns the array elements. In Lesson 11, "Changing Parameter Values," you learned that unless your programs pass parameters to a function by address, the function cannot change the parameter. However, as you can see here, the *get_values* function changes the array parameter *numbers*. As you will learn in Lesson 22, "Understanding Pointers," C++ actually passes arrays to functions using a pointer (a memory address). Therefore, the function can change the array elements as desired.

WHAT YOU MUST KNOW

In this lesson you learned that your programs can store multiple values of the same type within an array. C++ programs make extensive use of arrays. In Lesson 19, "Understanding Character Strings," you will learn that C++ programs use arrays to hold character strings. Before you continue with Lesson 19, however, make sure that you have learned the following key concepts:

- ☑ An array is a variable that can store one or more values of the same type.
- ☑ To declare an array you must specify a type, array name, and the number of values the array is to store.
- ☑ Values your programs store within an array are array elements.
- ☑ The first array element is element zero (*array[0]*); a value one less than the array size indexes the last array element.
- ☑ Programs often use index variables to access array elements.
- ☑ When a function receives an array as a parameter, the function must specify the array type and name, but not the array size.
- ☑ When a program passes an array to a function, the program will normally pass a parameter that tells the function the number of elements the array contains.
- ☑ Because C++ passes arrays to functions using the array's address, functions can change the values that the array contains.

LESSON 19

UNDERSTANDING CHARACTER STRINGS

Within C++ programs, character strings store information, such as filenames, book titles, employee names, and other character combinations. Most C++ programs make extensive use of character strings. As you will learn in this lesson, C++ stores character strings in an array of type *char*. You will learn how to store and manipulate character strings and how to take advantage of run-time library functions that manipulate character strings. By the time you finish this lesson, you will understand the following key concepts:

- To declare a character string, you must declare an array of type *char*.

- To assign characters to a character string, your programs assign characters to elements of the character-string array.

- C++ programs use the NULL (ASCII 0) character to mark the last character in the string.

- C++ lets your programs initialize character strings when the program declares the character-string variable.

- Programs can pass character strings to functions just as they would any array.

- Most C++ run-time libraries provide a set of functions that manipulate character strings.

C++ programs store character strings as an array of type *char*. Most C++ programs will make extensive use of character strings. Experiment with each program this lesson presents to ensure that you feel comfortable with strings. As you will find, working with character strings is similar to working with other arrays as discussed in Lesson 18, "Storing Multiple Values in Arrays."

DECLARING A CHARACTER STRING WITHIN YOUR PROGRAM

C++ programs make extensive use of character strings to store user names, filenames, and other character-based information. To declare a character string within your programs, you simply declare an array of type *char* with enough elements to hold the desired characters. For example, the following declaration creates a character-string variable named *filename* capable of storing 64 characters (do not forget that the NULL character, which you use to mark the end of the string, is one of these 64):

```
char filename[64];
```

As Figure 19.1 shows, this declaration creates an array indexed from *filename[0]* through *filename[63]*.

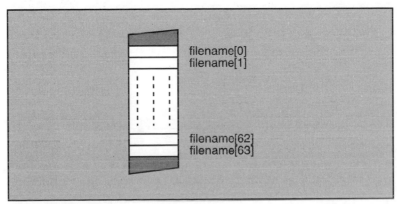

Figure 19.1 C++ *treats character strings as arrays of type* **char**.

127

The primary difference between character strings and other types of arrays is how C++ indicates the last element in the array. As you will learn, C++ programs represent the end of a character string using a NULL character, which C++ represents as the special character '\0'. When you assign characters to a character string, you must place the NULL ('\0') character after the last character in the string. For example, the following program, *Alphabet.CPP*, assigns the letters A through Z to the variable *alphabet* using a *for* loop. The program then appends the NULL character to the variable and displays the variable using *cout*:

```
#include <iostream.h>

void main(void)
 {
   char alphabet[27];  // 26 letters plus NULL
   char letter;
   int index;

   for (letter = 'A', index = 0; letter <= 'Z'; letter++, index++)
     alphabet[index] = letter;

   alphabet[index] = NULL;

   cout << "The letters are " << alphabet;
 }
```

As you can see, the program assigns the NULL character to the string to specify the string's last character:

```
alphabet[index] = NULL;
```

When the *cout* output stream displays the character string, it will display the string's characters one at a time until it encounters the NULL character. In short, the NULL character tells the program (or run-time library functions) the location of the last character in the string.

Take a close look at the *for* loop that appears in the previous program. As you can see, the loop initializes and increments two variables (*letter* and *index*). When a *for* loop initializes or increments more than one variable, you separate the operations using the C++ *comma* operator:

```
for (letter = 'A', index = 0; letter <= 'Z'; letter++, index++)
```

C++ AUTOMATICALLY APPENDS NULL TO STRING CONSTANTS

All the programs you have created throughout this book have used character string constants contained within double quotes as shown here:

```
"This is a string constant"
```

When you create a character string constant, the C++ compiler automatically assigns the NULL character, as shown in Figure 19.2.

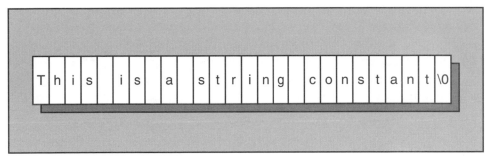

Figure 19.2 *The C++ compiler automatically appends the NULL character to string constants.*

When your programs display character string constants using the *cout* output stream, *cout* uses the NULL character, which the compiler appends to the string, to determine the last character it is to display.

Using the NULL Character

A character string is an array of characters followed by a NULL character ('\0'). When you declare a character string, you declare an array of type *char*. When the program later assigns characters to the string, the program is responsible for appending the NULL character, which represents the end of the string.

When you use string constants contained within double quotes, the C++ compiler automatically appends the NULL character. Most C++ functions use the NULL character to determine the string's last character.

The following program, *LoopNull.CPP*, changes the previous program slightly to use a *for* loop to display the string's contents:

```cpp
#include <iostream.h>

void main(void)
  {
    char alphabet[27];   // 26 characters plus NULL
    char letter;
    int index;

    for (letter = 'A', index = 0; letter <= 'Z'; letter++, index++)
      alphabet[index] = letter;

    alphabet[index] = '\0';

    for (index = 0; alphabet[index] != '\0'; index++)
      cout << alphabet[index];

    cout << endl;
  }
```

As you can see, the *for* loop examines the string's characters one at a time. If the character is not NULL (which indicates the string's last character), the loop will display the character, increment the index, and the process will continue. **129**

HOW 'A' DIFFERS FROM "A"

As you examine C++ programs, you will encounter characters contained within single quotes (such as 'A') and characters contained within double quotes ("A"). A character within single quotes is a *character constant*. The C++ compiler allocates only one byte of memory to store a character constant. A character in double quotes, however, contains a *string constant*, the character and a NULL character (which the compiler appends). Therefore, the compiler will allocate two bytes to store a character string. Figure 19.3 illustrates how the C++ compiler will store the character constant 'A' and the string constant "A".

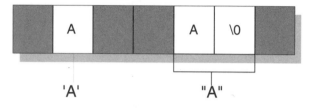

Figure 19.3 How the C++ compiler stores the character constant 'A' and the string constant "A".

INITIALIZING A CHARACTER STRING

As you learned in Lesson 18, C++ lets you initialize arrays at declaration. C++ character strings are no exception. To initialize a character string at declaration, you specify the string's characters within double quotes, as shown here:

```
char title[64] = "Rescued by C++";
```

If the number of characters you assign to the string is less than the array size, most C++ compilers will assign NULL characters to the remaining array elements. As was the case with arrays, if you do not specify the size of an character-string array that you initialize at declaration, the C++ compiler will allocate enough memory to hold the letters specified plus the NULL character. For example, the following statement creates a character string named *title* that contains the letters specified, including NULL:

```
char title[] = "Rescued by C++";
```

The following program, *Init_Str.CPP*, initializes a character string at declaration:

```
#include <iostream.h>

void main(void)
  {
    char title[64] = "Rescued by C++";
    char lesson[64] = "Understanding Character Strings";

    cout << "Book: " << title << endl;
    cout << "Lesson: " << lesson << endl;
  }
```

Several of the remaining programs this book presents initialize character strings in this way. Take time now to experiment with the *Init_Str.CPP* program by changing the characters the program assigns to each string.

PASSING STRINGS TO FUNCTIONS

Passing a character string to a function is very similar to passing any array as a parameter. Within the function, you simply specify the array type (*char*) and the left-and-right array brackets. You do not have to specify the string size. The following program, *Show_Str.CPP*, for example, uses the function *show_string* to display a character string to the screen:

```
#include <iostream.h>

void show_string(char string[])
  {
    cout << string << endl;
  }

void main(void)
  {
    show_string("Hello, C++!");
    show_string("I've been Rescued by C++");
  }
```

As you can see, the function *show_string* treats the character string parameter as an array, as shown here:

```
void show_string(char string[])
```

Because the NULL character indicates the end of the string, the function does not require a parameter that specifies the number of elements in the array. Instead, the function can determine the last element by searching the array for the NULL character. For example, the following program, *ByChar.CPP*, passes a character string to the *display_to_NULL* function, which in turn displays the string's letters one at a time until it encounters the NULL character:

```
#include <iostream.h>

void display_to_NULL(char string[])
  {
    for (int i = 0; string[i] != '\0'; i++)
       cout << string[i];
  }

void main(void)
  {
    display_to_NULL("Rescued by C++");
  }
```

As you have learned, C++ functions often use the NULL character to determine the end of a string.

The following program, *Str_Len.CPP*, creates a function named *string_length*, which searches a string for the NULL character to determine the number of characters the string contains. The function then uses a *return* statement to return the string length to the caller. The *Str_Len.CPP* program passes several different character strings to the function, displaying the length of each string to the screen:

```
#include <iostream.h>

int string_length(char string[])
  {
```

131

```
    int i;

    for (i = 0; string[i] != '\0'; i++);  // Do nothing but
                                           // loop to next
                                           // character


    return(i);  // The length of the string
 }

 void main(void)
  {
    char title[] = "Rescued by C++";
    char lesson[] = "Understanding Character Strings";

    cout << "The string " << title << " contains " <<
      string_length(title) << " characters" << endl;

    cout << "The string " << lesson << " contains " <<
      string_length(lesson) << " characters" << endl;
  }
```

As you can see, the function starts with the string's first character (element 0) and then examines each element until it encounters the NULL. As you examine more C++ programs, you will encounter many functions that search character strings for the NULL character in this way.

TAKE ADVANTAGE OF THE FACT THAT NULL IS AN ASCII 0

As you have learned, the NULL character is the ASCII 0 character. In Lesson 8, "Teaching Your Program to Make Decisions," you learned that C++ uses the value 0 to represent false. Therefore, because the NULL character is equal to 0, your programs can simplify many loop operations. For example, many functions search character strings, character by character in search of NULL. The following *for* loop illustrates how a program might search a string for NULL:

```
for (index = 0; string[index] != NULL; index++)
  ;
```

Because the NULL character is equal to 0, many programs simplify loops that search for NULL as shown here:

```
for (index = 0; string[index]; index++)
  ;
```

In this example, as long as the character contained in *string[index]* is not NULL (0 or false), the loop will continue.

USING RUN-TIME LIBRARY STRING FUNCTIONS

In Lesson 12, "Taking Advantage of the Run-Time Library," you learned that most C++ compilers provide a very powerful collection of functions called the run-time library. When you examine the run-time library, you will find that it contains many different string-manipulation functions. For example, the function *strupr* converts a character string to uppercase. Likewise, the function *strlen* returns the number of characters in a string. Most run-time libraries even provide functions that let you search strings for specific characters. The following program, *StrUpr.CPP*, for example, illustrates the use of *strupr* and *strlwr* run-time library functions:

```
#include <iostream.h>
#include <string.h> // Contains function prototypes

void main(void)
 {
   char title[] = "Rescued by C++";
   char lesson[] = "Understanding Character Strings";

   cout << "Uppercase: " << strupr(title) << endl;
   cout << "Lowercase: " << strlwr(lesson) << endl;
 }
```

Using the run-time library string-manipulation functions can save you considerable programming. Take time now to print a copy of the header file *string.h* to determine string manipulation routines your compiler's run-time library supports.

You Must Play by the Rules

As you have learned, most functions that manipulate strings rely on the NULL character to determine the string's last character. When your programs assign strings to characters, your programs must insure that they append the NULL character as the last character in the string. If your programs do not use the NULL in a consistent way, functions that rely on the NULL character will fail.

WHAT YOU MUST KNOW

Most C++ programs make extensive use of character strings. In this lesson you learned how to work with strings within your programs. In Lesson 20, "Storing Related Information in Structures," you will learn how to store related information of different types in a C++ structure variable. Using a structure, you can store such information as an employee's name, age, salary, and phone number, all within one variable. Before you continue with Lesson 20, however, make sure that you have learned the following key concepts:

☑ A character string is an array of characters that programs terminate with the ASCII 0 (NULL) character.

☑ You create a character string within your programs by declaring an array of type *char*.

☑ Your program is responsible for placing the NULL character after the string's last character.

☑ When your program uses string constants contained within double quotes, the C++ compiler automatically appends the NULL character.

☑ C++ lets you initialize strings at declaration by specifying the characters within double quotes.

☑ Most C++ compilers provide an extensive set of string-manipulation functions in their run-time library.

LESSON 20

STORING RELATED INFORMATION IN STRUCTURES

In Lesson 18, "Storing Multiple Values in Arrays," you learned that C++ lets you store related information of the same type in arrays. As you found, grouping related values in arrays is very convenient. As your programs become more complex, your programs may need to group related information that may not be the same type. For example, assume your program works with employee records. Your program must track an employee's name, age, salary, address, employee number, position number, and so on. To store this information, your program will need variables of type *char*, *int*, *float*, as well as character strings.

When your program must store related information of different types, your program can use a *structure*. As you will learn, a structure is a variable that groups related pieces of information, called *members*, whose types may differ. By grouping data into one variable in this way, you simplify your programs by reducing the number of variables you must manage, pass to functions, and so on. This lesson examines how your programs create and use structures. By the time you finish this lesson, you will understand the following key concepts:

- Structures let your programs group related data, whose types may differ, into one variable.

- A structure consists of one or more pieces of data, called *members*.

- To define a structure within your program you specify the structure name and its members.

- Each structure member has a type, such as *char*, *int*, and *float*, and each member name must be unique to the structure.

- After your program defines a structure, your program can declare variables of the structure type.

- To change structure members within a function, your programs must pass the structure to the function by address.

In Section 4 of this book, you will examine C++ object-oriented programming. Your ability to understand and work with structures will make using C++ object-oriented classes much easier. Take time to experiment with the programs this lesson presents and make sure you feel comfortable with the process of assigning and retrieving values from structure members.

DECLARING A STRUCTURE

A structure defines a *template* that your program can later use to declare one or more variables. In other words, your program first defines a structure and then declares variables of the structure type. To define a structure, your programs use the *struct* keyword, usually followed by a name and left brace. Following the open brace, you specify the type and name for one or more *members*. After the last member, you place a closing right brace. At that time, you can optionally declare variables of the structure type, as shown here:

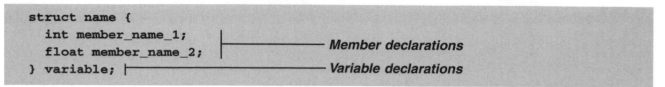

```
struct name {
  int member_name_1;
  float member_name_2;                  Member declarations
} variable;                             Variable declarations
```

134   For example, the following definition creates a structure named *employee* that holds employee information:

```
struct employee {
  char name[64];
  long employee_id;
  float salary;
  char phone[10];
  int office_number;
};
```

In this example, the structure definition does not declare any variables of the structure type. After you define a structure, your program can declare variables of the structure type using the structure name (sometimes called the *structure tag*), as shown here:

```
employee boss, worker, new_employee;
```
— Tag
— Variable declarations

In this example, the statement creates three variables of the *employee* structure. As you examine C++ programs, you may find declarations that precede the structure tag with the keyword *struct*, as shown here:

```
struct employee boss, worker, new_employee;
```

C programming requires the *struct* keyword, so some programmers might include it by habit. In C++, however, the use of the *struct* keyword is optional.

USING STRUCTURE MEMBERS

The structure lets your program group information, called members, into one variable. To assign a value to a member or to access a member's value, you use the C++ *dot operator* (.). For example, the following statements assign values to different members of a variable named *worker* of the type *employee*:

```
worker.employee_id = 12345;
worker.salary = 25000.00;
worker.office_number = 102;
```

To access a structure member, you specify the variable name, followed by a dot operator and member name (*structure_name.member*). For example, the following statements assign values stored in different structure members to program variables:

```
identification = worker.employee_id;
salary = worker.salary;
office = worker.office_number;
```

The following program, *Employee.CPP*, illustrates the use of a structure of type *employee*:

```
#include <iostream.h>
#include <string.h>

void main(void)
  {
    struct employee {
      char name[64];
      long employee_id;
      float salary;
      char phone[10];
```

135

```
      int office_number;
    } worker;

    // Copy a name to the string
    strcpy(worker.name, "John Doe");

    worker.employee_id = 12345;
    worker.salary = 25000.00;
    worker.office_number = 102;

    // Copy a phone number to the string
    strcpy(worker.phone, "555-1212");

    cout << "Employee: " << worker.name << endl;
    cout << "Phone: " << worker.phone << endl;
    cout << "Employee id: " << worker.employee_id << endl;
    cout << "Salary: " << worker.salary << endl;
    cout << "Office: " << worker.office_number << endl;
  }
```

As you can see, the program can assign values to the structure's integer and floating-point members in a straight-forward way. The program simply uses the assignment operator to assign a value to the corresponding member. Note, however, the use of the *strcpy* function to copy character strings to the members *name* and *phone*. Unless you initialize the structure members explicitly when you declare the structure variable, you must copy character strings to character-string members. Later in this lesson, you will learn how to initialize members when you declare a structure variable.

Declaring Structure Variables

C++ structures let your programs group related information, whose types may differ, into one variable. A structure defines a template for your program's future variable declarations. Each structure has a unique name (sometimes called a *tag*). Using the structure name, you can declare variables of the structure type. Programmers call the information the structure stores members. To use or assign a member value, you use the C++ dot operator, as shown here:

```
variable.member = some_value;
some_variable = variable.other_member;
```

STRUCTURES AND FUNCTIONS

If a function does not change a structure, you can pass the structure to the function by name. For example, the following program, *Show_Emp.CPP*, uses the function *show_employee* to display the members of a structure of type *employee*:

```
#include <iostream.h>
#include <string.h>

struct employee {
  char name[64];
  long employee_id;
```

```
    float salary;
    char phone[10];
    int office_number;
  };

void show_employee(employee worker)
  {
    cout << "Employee: " << worker.name << endl;
    cout << "Phone: " << worker.phone << endl;
    cout << "Employee id: " << worker.employee_id << endl;
    cout << "Salary: " << worker.salary << endl;
    cout << "Office: " << worker.office_number << endl;
  }

void main(void)
  {
    employee worker;

    // Copy a name to the string
    strcpy(worker.name, "John Doe");

    worker.employee_id = 12345;
    worker.salary = 25000.00;
    worker.office_number = 102;

    // Copy a phone number to the string
    strcpy(worker.phone, "555-1212");
    show_employee(worker);
  }
```

As you can see, the program passes the structure variable *worker*, by name, to the function *show_employee*. Then, within the function, *show_employee* displays the structure members. Note, however, that the program now defines the structure *employee* outside of *main* and before the function *show_employee*. Because the function declares the variable *worker* as type *employee*, the definition of the *employee* structure must precede the function.

FUNCTIONS THAT CHANGE STRUCTURE MEMBERS

As you have learned, when a function changes a parameter, you must pass the parameter to the function by address. If a function must change a structure member's value, you must pass the structure to the function by address. To pass a structure variable by address, you simply precede the variable name with the C++ address operator (&), as shown here:

```
some_function(&worker);
```

Within a function that changes one or more members, you must work with a pointer. When you use a pointer to a structure, the easiest way to reference a structure member is to use the following syntax:

```
pointer_variable->member = some_value;
```

The following program, *Chg_Mbr.CPP*, for example, passes a structure of type *employee* to the *get_employee_id* function, which prompts the user for an employee identification number and then assigns the number to the structure's *employee_id* member. To change the member, the function works with a pointer to the structure:

```
#include <iostream.h>
#include <string.h>

struct employee {
  char name[64];
  long employee_id;
  float salary;
  char phone[10];
  int office_number;
};

void get_employee_id(employee *worker)
 {
   cout << "Type in an employee id: ";
   cin >> worker->employee_id;
 }

void main(void)
 {
   employee worker;

   // Copy a name to the string
   strcpy(worker.name, "John Doe");
   get_employee_id(&worker);

   cout << "Employee: " << worker.name << endl;
   cout << "Id: " << worker.employee_id << endl;
 }
```

As you can see, within *main* the program passes the structure variable *worker* to the function *get_employee_id* by address. Within the function, *get_employee_id* assigns the value the user types to the *employee_id* member using the following statement:

```
cin >> worker->employee_id;
```

Working with Pointers to Structures

When a function changes a structure member, the caller must pass the structure to the function by address. The function, in turn, uses a pointer to a structure. To access a structure member, the function should use the following format:

```
value = variable->member;
variable->other_member = some_value;
```

INITIALIZING STRUCTURE MEMBERS

When you declare structure variables, C++ lets you initialize the variable's members. For example, the following declaration creates and initializes a structure variable of type *book*:

```
struct book {
   char *title;
   float price;
   char *author;
} computer_book = { "Rescued by C++", 29.95, "Jamsa" };
```

As you can see, the statement includes the variable's initial values within the left-and-right braces, much like the technique you used to initialize array elements in Lesson 18.

WHAT YOU MUST KNOW

Structures let your programs group related pieces of information, whose types may differ, into the same variable. By combining data into one variable in this way, your programs can better represent objects that consist of two or more pieces, such as employees, books, and so on. In Lesson 21, "Understanding Unions," you will learn how to use C++ unions, which, like structures, use members. However, you will learn that unions use memory much differently than structures. A union can store only one value at any given time, regardless of the number of members. Before you continue with Lesson 21, however, make sure that you have learned the following key concepts:

- ☑ Structures let your programs group related pieces of information, whose types may differ, into one variable.

- ☑ Programmers call the pieces of information that make up a structure members.

- ☑ A structure defines a template your programs can use to declare variables.

- ☑ After you define a structure, you can use the structure name (tag) to declare variables of the structure type.

- ☑ To assign a value to, or to access the value of, a structure member, your programs use the C++ dot operator, such as *variable.member*.

- ☑ If a function changes the value of a structure member, the program must pass the structure variable to the function by address.

- ☑ When a function uses a pointer to a structure, the function should use the format *variable->member* to access a structure member.

LESSON 21

UNDERSTANDING UNIONS

In Lesson 20, "Storing Related Information in Structures," you learned how to group related information into a single variable using C++ structures. As your programs become more complex, there may be times when your programs need different ways to view the same piece of information. In addition, there may be times when your programs must work with two or more values, but will only use one of the values at any time. In such cases, your programs can use *unions* to store the data. In this lesson, you will learn how to create and use unions to store information. As you will learn, unions are very similar to the structures you examined in Lesson 20. By the time you finish this lesson, you will understand the following key concepts:

- C++ unions are very similar to structures, with the exception of how C++ stores each in memory—a union can store a value for only one member at a time.

- A union is a data structure, which like a structure, consists of one or more members.

- A union defines a template with which your programs can later declare variables.

- To access a specific union member, your programs use the C++ dot operator.

- To change the value of a union member within a function, your program must pass the union variable to the function by address.

- An *anonymous union* is a union that does not have a name (a tag).

As you will learn, unions are similar to C++ structures, however, the way C++ stores unions is very different from the way it stores structures.

How C++ Stores Unions

Within your program, a C++ union is very similar to a structure. For example, the following statement defines a union, named *distance*, that contains two members:

```
union distance {
  int miles;
  long meters;
};
```

As was the case with a structure, the union definition does not allocate memory. Instead, the definition provides a template with which your programs can later declare a union variable. To declare a union variable, you can use either of the following formats:

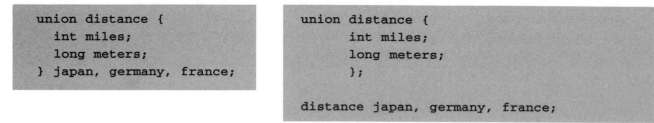

```
union distance {
  int miles;
  long meters;
} japan, germany, france;
```

```
union distance {
  int miles;
  long meters;
};

distance japan, germany, france;
```

As you can see, the *distance* union contains two members, *miles* and *meters*. The declarations create variables that let you store the distance to the specified countries. Like a structure, your program can assign a value to either member. However, unlike a structure, your program can only assign a value to one member at a time. In other words, when your program assigns a value to one union member, the new value overwrites the value your program may have previously assigned to the same or different member within the union. When you declare a union, the C++ compiler allocates memory to store the union's largest member. In the case of the *distance* union, the compiler allocates enough memory to store a value of type *long*, as Figure 21.1 shows.

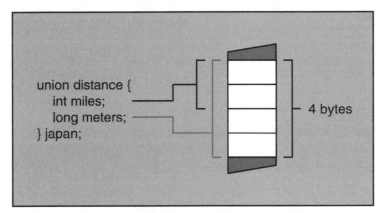

Figure 21.1 C++ allocates only enough memory to hold a union's largest member.

Assume that your program assigns a value to the *miles* member, as shown here:

```
japan.miles = 12123;
```

If your program later assigns a value to the *meters* member, the new value overwrites the value assigned to the *miles* member.

The following program, *UseUnion.CPP*, illustrates the use of the *distance* union. The program first assigns a value to the *miles* member and then displays the value. The program then assigns a value to the *meters* member. At that time, by assigning a value to the *meters* member, the program overwrites the *miles* member's value, as shown here:

```cpp
#include <iostream.h>

void main(void)
  {
    union distance {
      int miles;
      long meters;
    } walk;

    walk.miles = 5;

    cout << "A distance walked in miles is " << walk.miles << endl;

    walk.meters = 10000;

    cout << "A distance walked in meters is " << walk.meters << endl;
  }
```

As you can see, the program accesses union members using the dot notation similar to that you used to access structure members in Lesson 20.

Unions Hold Only One Member's Value at a Time

A union is a data structure, which like a structure, lets your programs store related pieces of information within one variable. Unlike a structure, however, a union holds only one member's value at a given time. In other words, when you assign a value to a union member, you overwrite any previous assignment. A union defines a template with which your programs can later declare variables. When the C++ compiler encounters a union definition, the compiler allocates only enough memory to hold the largest member in the union.

UNDERSTANDING C++ ANONYMOUS UNIONS

An *anonymous union* is a union that does not have a name. C++ provides anonymous unions to simplify the use of union members in programs that are using a union to save memory or to create an alias for a value. For example, assume that your program requires two variables, *miles* and *meters*. Likewise, assume that the program only uses one of the variables at any given time. Using the *distance* union just discussed, the program would have to use the union members *name.miles* and *name.meters*. As an alternative, the following statement creates an anonymous (unnamed) union:

```
union {
   int miles;
   long meters;
};
```

As you can see, the declaration does not use a union name, and it does not declare a variable of the union type. The program in turn, can refer to the member names *miles* and *meters* without using the dot notation. The following program, *Anonym.CPP*, creates an anonymous union that contains the *miles* and *meters* members. As you can see, the program treats the members just as it would normal variables. The difference between the members and a normal variable, however, is that when you assign a value to either member, you overwrite the other member's value:

```
#include <iostream.h>

void main(void)
 {
   union {
      int miles;
      long meters;
   };

   miles = 10000;

   cout << "The value of miles is " << miles << endl;

   meters = 150000L;
   cout << "The value of meters is " << meters << endl;
 }
```

As you can see, using the anonymous union the program can save memory without the burden of using the union name and dot notation to access member values.

Anonymous Unions Let Your Programs Save Space

An anonymous union is an unnamed union. Anonymous unions provide your programs with a way to save memory without the burden of having to use the dot notation. The following statements define an anonymous union capable of storing two character strings:

```
union {
  char short_name[13];
  char long_name[255];
};
```

WHAT YOU MUST KNOW

In this lesson you learned how to create and use unions within your programs. As you have learned, the format of a union is very similar to that of a structure. However, the way C++ stores unions is much different than how it stores a structure. In Lesson 11, "Changing Parameter Values," you first learned that for a function to change a parameter, your programs had to pass the parameter to the function using a pointer (or memory address). Since Lesson 11, your programs have used pointers for arrays and character strings. In Lesson 22, "Understanding Pointers," you will take another look at C++ pointer operations. Before you continue with Lesson 22, however, make sure that you have learned the following key concepts:

- ☑ When you declare a union, the C++ compiler allocates only enough memory to hold the union's largest member.

- ☑ A union definition does not allocate memory, instead it provides a template with which your programs can later declare variables.

- ☑ Programs access union members using dot notation. When your program assigns a value to a union member, your program overwrites a value previously assigned to a different union member. In other words, a union can hold a value for only one member at a time.

- ☑ An anonymous union is a union that does not have a name. When a program declares an anonymous union, the program can use the union's members like any variable without the burden of dot notation.

LESSON 22

UNDERSTANDING POINTERS

As you have learned, C++ programs store variables in memory. A pointer is a memory address that "points to" or references a specific location. In Lesson 11, "Changing Parameter Values," you learned that to change a parameter within a function, your programs must pass the parameter's address (a pointer) to the function. The function, in turn, uses a pointer variable to access the corresponding memory location. Several of the programs you have created in the past few lessons have used pointers to parameters. Likewise, when your programs work with character strings and arrays, it is common for your programs to use pointers to manipulate the array elements. Because the use of pointers is so common, it is very important that you readily understand their use. Therefore, this lesson takes another look at pointers. By the time you finish this lesson, you will understand the following key concepts:

- For simplicity (to reduce code), many programs treat a character string as a pointer and manipulate the string's contents using pointer operations.

- When you increment a pointer variable (a variable that stores an address), C++ automatically increments the address by the correct amount (1 byte for *char*, 2 bytes for *int*, four bytes for *float*, and so on), so that the pointer points to the next value of the pointer's type.

- Your programs can use pointers to work with an array of integer or floating-point values.

Pointer operations are very common in C++. Take time to experiment with the programs this lesson presents.

USING A POINTER TO A CHARACTER STRING

As you have learned, a pointer contains a memory address. When your programs pass arrays (for example, a character string) to a function, C++ passes the address of the array's first element as the parameter. As a result, it is very common for functions to use a pointer to a character string. To declare a pointer to a character string, the function simply precedes the variable name with an asterisk, as shown here:

```
void some_function(char *string);
```

The asterisk that precedes the variable name tells C++ the variable will store a memory address, a pointer. The following program, *Ptr_Str.CPP*, uses a pointer to a character string within the function *show_string* to display the string's contents one character at a time:

```
#include <iostream.h>

void show_string(char *string)
  {
    while (*string != '\0')
      {
        cout << *string;
        string++;
      }
  }
```

```
void main(void)
{
   show_string("Rescued By C++!");
}
```

Take a close look at the *while* loop within the *show_string* function. The condition *while (*string != '\0')* tests whether the current letter the *string* pointer points to is not equal to NULL, which indicates the string's last character. If the character is not NULL, the loop will display the current character using *cout*. Next, the statement *string++;* increments the *string* pointer so that it points to the next character in the string. When the *string* pointer points to the NULL character, the function has displayed the string and the loop ends.

Assume, for example, the string passed to the function resides at the address 1000 within your computer's memory. Each time the function increments the *string* pointer, the pointer points to the next character (address 1001, 1002, 1003, and so on), as shown in Figure 22.

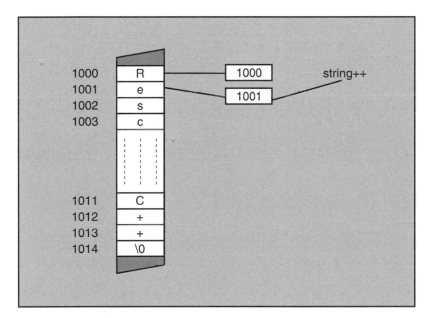

Figure 22 Traversing a string using a pointer.

LOOKING AT A SECOND EXAMPLE

As you just learned, by using a pointer your functions can traverse a string's characters until they find the NULL character. The following program, *Ptr_Len.CPP*, uses a string pointer within the *string_length* function to determine the number of characters in a string:

```
#include <iostream.h>

int string_length(char *string)
{
   int length = 0;
   while (*string != '\0')
   {
      length++;
```

```
      string++;
    }

  return(length);
 }

void main(void)
 {
   char title[] = "Rescued By C++";

   cout << title << " contains " << string_length(title) << " characters";
 }
```

As you can see, the *string_length* function loops through the string's characters until it finds the NULL character.

Incrementing a Character String Pointer

 When a program passes an array to a function, C++ passes the memory address of the array's first element. Using a pointer variable, a function can move through the array's contents by incrementing the pointer value. For example, assume a program passes the character string "Hello" to a function. Within the function, the pointer variable initially points to the memory location that contains the letter 'H'. When the function increments the pointer, the pointer then points to the memory location that contains the letter 'e'. As the function continues to increment the pointer's value, the pointer will point at each letter in the string and, finally, the NULL character.

TRIMMING EXCESS STATEMENTS

To locate the end of a character string, each of the previous programs used the following *while* loop:

```
while (*string != '\0')
```

As discussed, the NULL character ('\0') is the ASCII value 0. Because C++ uses the value 0 to represent false, your programs can write the previous loop as follows:

```
while (*string)
```

In this example, as long as the character the string pointer points to is not 0 (NULL), the condition evaluates as true, and the loop will continue.

In Lesson 6, "Performing Simple Operations," you learned that the C++ postfix increment operator lets you use a variable's value and then increments that value. Many C++ programs use the postfix increment and decrement operators to loop through arrays using pointers. For example, using the postfix increment operator, the following *while* loops are identical:

```
while (*string)
  {
    cout << *string;
    string++;
  }
```

```
while (*string)
  cout << *string++;
```

The statement *cout << *string++;* directs C++ to display the character that *string* points to and then to increment *string's* current value to point to the next character. Using these techniques, the following program, *SmartPtr.CPP*, illustrates new implementations of the *show_string* and *string_length* functions:

```cpp
#include <iostream.h>

void show_string(char *string)
 {
   while (*string)
     cout << *string++;
 }

int string_length(char *string)
 {
   int length = 0;

   while (*string++)
     length++;

   return(length);
 }

void main(void)
 {
   char title[] = "Rescued By C++";

   show_string(title);
   cout << " contains " <<
     string_length(title) << " characters";
 }
```

As you encounter C++ functions that manipulate strings using pointers, the functions will likely use these shorthand notations.

Traversing a Character String

One of the most common uses of pointers within C++ programs is to traverse character strings. To reduce programming, many programs will use the following statements to traverse a string:

```cpp
while (*string)
  {
      // statements
      string++;   // point to the next character
  }
```

The following function, *string_uppercase*, uses pointers to convert the characters in a string to uppercase:

```cpp
char *string_uppercase(char string)
  {
```

147

```
    char *starting_address = string;   // address of string[0];
    while (*string)
      {
        if ((*string >= 'a') && (*string <= 'z'))
          *string = *string - 'a' + 'A';
        string++;
      }
    return(starting_address);
  }
```

The *string_uppercase* function stores and returns the string's starting address, which lets your programs use the function as follows:

```
  cout << string_uppercase("Hello, world!") << endl;
```

By returning the string's starting address to the caller, the code can use the function's return address to access the string. In this example, the code uses the string's address within the *cout* stream to display the string's contents.

USING POINTERS WITH OTHER ARRAY TYPES

Although the use of pointers is most common with character strings, you can use pointers with other array types. For example, the following program, *PtrFloat.CPP*, uses a pointer to an array of type *float* to display floating-point values:

```
#include <iostream.h>

void show_float(float *array,
  int number_of_elements)
 {
   int i;

   for (i = 0; i < number_of_elements; i++)
     cout << *array++ << endl;
 }

void main(void)
 {
   float values[5] = {1.1, 2.2, 3.3, 4.4, 5.5};

   show_float(values, 5);
 }
```

As you can see, within the *show_float* function the *for* loop uses the value the *array* pointer points to and then increments the pointer's value to point to the next value. In this example, the program has to pass a parameter that specifies the number of array elements because, unlike character strings, arrays of type *float* (or *int*, *long*, and so on), do not use the NULL character to indicate the last element.

UNDERSTANDING POINTER MATH

As you have learned, your programs can use pointers to arrays of any type. In the previous program, for example, the *show_float* function incremented a pointer to move through an array of type *float*. As you have learned, a pointer points to a memory location that contains a value of a specific type, such as *char*, *int*, or *float*. When a function traverses an array using a pointer, the function increments the pointer to move from one value to the next. For a pointer to point to the next value in the array, C++ must keep track of each value's size (in bytes), so it knows by how much it must increment the pointer variable. For example, to move the pointer to the next character in an array, C++ must increment the pointer value by one byte. However, to point to the next value in an array of type *int*, C++ must increment the pointer by two bytes (values of type *int* require two bytes of memory). For values of type *float*, C++ increments the pointer by four bytes. By knowing the type of value to which the pointer points, C++ knows by how much it must increment the pointer value. Within your programs, you simply use an increment operator, such as *pointer++*, to increment a pointer variable. Behind the scenes, however, C++ increments the actual value the pointer contains (the memory address) by the correct amount.

WHAT YOU MUST KNOW

C++ programs make extensive use of pointers, particularly to manipulate character strings. Because programmers commonly use pointer operations, this lesson took a second look at pointer operations in C++. In Lesson 23, "Starting with C++ Classes," you will start using C++ object-oriented capabilities. To begin, you will create classes, which are similar to structures. Your programs will use a class to define an object, such as *file*. Within the class, you will specify the functions that manipulate the object, such as *print_file* or *delete_file* functions. Before you continue with Lesson 23, however, make sure that you have learned the following key concepts:

- ☑ Pointers contain a memory address. When you pass an array to a function, C++ passes the address of the first element in the array.

- ☑ By incrementing a pointer's value, you can direct the pointer to point to the next value in an array.

- ☑ Functions that manipulate strings using pointers normally traverse the string until the pointer points to the NULL character.

- ☑ When you use pointers to other array types, your functions must know the number of elements in the array or a special ending value.

- ☑ When you use pointers to other array types, C++ automatically (behind the scenes) increments the pointer (the memory address) by the correct amount so the pointer points to the next value in the array.

CONCENTRATED C++

Abstract:

A basic introduction to Object Oriented Programming with C++. An ability to design and write programs in a modern procedural language like Ada, Modula-2 or Pascal is assumed. C experience is not required. The coverage of the language is comprehensive. Classes, single and multiple inheritance, virtual functions and templates are discussed. However, exception handling is not described. Compiler specific features are not used. This material is suitable for use with any quality C++ compiler.

CONCENTRATED C++
An Introduction to the Language (2nd Edition)

Adrian P. Robson
The University of Northumbria at Newcastle

http://cmwww.unn.ac.uk/~adrian/c++/cpppaper/—cpppaper.html

BORLAND C++ BOOKS

BorlandOnline   Products | Search | Downloads | Membership | Newsgroups | Feedback

Borland C++

Borland C++

IS SUPPORTED BY THESE TITLES FROM LEADING BOOK PUBLISHERS

- Borland C++ 5.0
- Borland C++ 4.5
- Borland C++ 4.0
- Borland C++ 3.1
- Turbo C++ 4.5 for Windows
- Turbo C++ Visual Edition for Windows
- Turbo C++ 3.0
- Turbo Assembler/Turbo Debugger

Borland C++ 5.0
Borland C++ 5 Resource Kit by Borland Press

http://www.borland.com/borlandcpp/books/index.html

THE C++ VIRTUAL LIBRARY

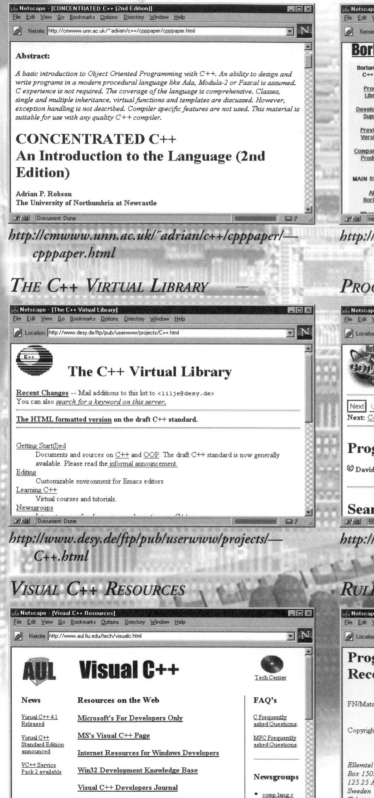

The C++ Virtual Library

Recent Changes -- Mail additions to this list to <liljie@desy.de>
You can also *search for a keyword on this server.*

The HTML formatted version on the draft C++ standard.

Getting Start(l)ed
    Documents and sources on C++ and OOP. The draft C++ standard is now generally available. Please read the informal announcement.
Editing
    Customizable environment for Emacs editors
Learning C++
    Virtual courses and tutorials.
Newsgroups

http://www.desy.de/ftp/pub/userwww/projects/—C++.html

PROGRAMMING IN C

Next | Up | Previous
Next: Copyright

Programming in C

© David Marshall 1994

Search for Keywords in C Notes

http://www.cm.cf.ac.uk/Dave/C/CE.html

VISUAL C++ RESOURCES

AUL Visual C++   Tech Center

News	Resources on the Web	FAQ's
Visual C++ 4.1 Released	**Microsoft's For Developers Only**	C Frequently asked Questions.
Visual C++ Standard Edition announced	**MS's Visual C++ Page**	MFC Frequently asked Questions.
VC++ Service Pack 2 available	**Internet Resources for Windows Developers**	
	Win32 Development Knowledge Base	**Newsgroups**
	Visual C++ Developers Journal	• comp.lang.c • comp.lang.c++ • comp.std.c

http://www.aul.fiu.edu/tech/visualc.html

RULES AND RECOMMENDATIONS

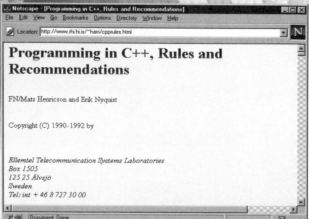

Programming in C++, Rules and Recommendations

FN/Mats Henricson and Erik Nyquist

Copyright (C) 1990-1992 by

Ellemtel Telecommunication Systems Laboratories
Box 1505
125 25 Älvsjö
Sweden
Tel: int + 46 8 727 30 00

http://www.rhi.hi.is/~harri/cpprules.html

USING C++ CLASSES

Object-oriented programming focuses on the objects (things) that make up a system. For example, you might have a file object, or an employee object, and so on. Each object stores related data, such as a file or employee name, an employee identification number, and possibly a salary. As such, objects are similar to C++ structures. However, objects also define a set of operations your programs perform on the object's data. In the case of a file object, for example, you might print, delete, or even copy a file. Likewise, given an employee object, you might print, promote, or possibly even fire the object. C++ uses a class to store an object's data and the functions that operate on the data. This section examines C++ classes in detail. The lessons in this section include:

LESSON 23

STARTING WITH C++ CLASSES

The class is the primary C++ tool for object-oriented programming. As you will learn in this lesson, a class is very similar to a structure in that it groups members that correspond to an object's data, as well as functions (called *methods*) that operate on the data. As you will learn, an object is a *thing*, such as a telephone, or a file, or a book. A C++ class lets your programs define the object's attributes (characteristics). In the case of a *telephone* object, the class might contain data members, such as the phone's number and type (tone or rotary), and functions that operate on the phone, such as *dial*, *answer*, and *hang_up*. By grouping an object's data and code into one variable, you simplify programming and increase your code reuse (your ability to use the code in a second unrelated program). This lesson introduces C++ classes. By the time you finish this lesson, you will understand the following key concepts:

- To define a class your program must specify the class name, the class data members, and the class functions (methods).

- A class definition provides a template which your programs use to create objects of that class type, much like your programs create variables from the types *int*, *char*, and so on.

- Your programs assign values to class-data members using the dot operator.

- Your programs invoke class-member functions (*methods*) using the dot operator.

UNDERSTANDING OBJECTS AND OBJECT-ORIENTED PROGRAMMING

In the simplest sense, an *object* is a thing. When you create a program, the program normally uses variables to store information about different real-world things, such as employees, books, and even files. When you perform object-oriented programming, you focus on the things that make up a system and the operations you must perform on those things. For example, given a file object, you might have operations that print, display, delete, or change the file. In C++, you use a class to define your objects. When you define a class, your goal is to include as much information about the object within the class as you can. You can therefore "pick up" a class that you create for one program and use it in many different programs.

A C++ class lets your program group data and functions that perform operations on the data. Most books and articles on object-oriented programming refer to the class functions as *methods*. Like a structure, a C++ class must have a unique name, followed by an opening brace, one or more members, and a closing brace, as shown here:

```
class class_name {
   int data_member;          // Data member
   void show_member(int);    // Function member
};
```

After you define a class, you can declare variables of the class type (called *objects*) as shown here:

```
class_name object_one, object_two, object_three;
```

The following definition creates an *employee* class that contains data variables and method definitions:

```
class employee {
 public:
   char name[64];
   long employee_id;
   float salary;
   void show_employee(void)
    {
      cout << "Name: " << name << endl;
      cout << "Id: " << employee_id << endl;
      cout << "Salary: " << salary << endl;
    };
};
```

In this example, the *employee* class contains three variable members and one function member. Note the use of the *public* label within the class definition. As you will learn in Lesson 24, "Understanding Private and Public Data," class members can be *private* or *public,* which controls how your programs can access the members. In this example all the members are *public,* which means your program can access any member using the dot operator. After you define the class within your program, you can declare objects (variables) of the class type, as shown here:

```
employee worker, boss, secretary;          ── Class name
                                           ── Class variable (objects)
```

The following program, *EmpClass.CPP*, creates two *employee* objects. Using the dot operator, the program assigns values to the data members. The program then uses the *show_employee* member to display the employee information, as shown here:

```
#include <iostream.h>
#include <string.h>

class employee {
 public:
   char name[64];
   long employee_id;
   float salary;
   void show_employee(void)
    {
      cout << "Name: " << name << endl;
      cout << "Id: " << employee_id << endl;
      cout << "Salary: " << salary << endl;
    };
};

void main(void)
  {
    employee worker, boss;

    strcpy(worker.name, "John Doe");
    worker.employee_id = 12345;
    worker.salary = 25000;

    strcpy(boss.name, "Happy Jamsa");
    boss.employee_id = 101;
    boss.salary = 101101.00;
```

```
      worker.show_employee();
      boss.show_employee();
  }
```

As you can see, the program declares two *employee* objects, *worker* and *boss* and then uses the dot operator to assign values to the members and to invoke the *show_employee* function.

Understanding Objects

Most C++ programs represent real-world entities, or objects. In the simplest sense, an object is a thing, such as a car, a dog, a clock, and so on. An object normally has several attributes and operations that your program must perform on the attributes. For example, in the case of the clock object, the attributes might include the current time and an alarm time. Operations your program might perform on the clock include setting the time, setting the alarm, or turning the alarm off. When your programs perform object-oriented programming, your programs focus on objects and operations on those objects.

DECLARING CLASS METHODS OUTSIDE OF THE CLASS

In the previous *employee* class, the class defined the function within the class itself (called an *inline function*). As functions become long, defining the functions within the class can clutter the class definition. As an alternative, you can place a function prototype within the class and then define the function outside of the class. In this case, the class definition with the prototype becomes the following:

```
class employee {
  public:
    char name[64];
    long employee_id;
    float salary;
    void show_employee(void); |————————— Function prototype
};
```

Because different classes may use functions of the same name, you must precede function names you declare outside of the class with the class name and *global resolution operator* (::). In this case, the *show_employee* function definition becomes the following:

```
void employee::show_employee(void)                ——— Class name
  {                                               ——— Member name
    cout << "Name: " << name << endl;
    cout << "Id: " << employee_id << endl;
    cout << "Salary: " << salary << endl;
  };
```

As you can see, the code precedes the function definition with the class name (*employee*) and the global resolution operator (::). The following program *ClassFun.CPP*, moves the definition of the *show_employee* function outside of the class using the global resolution operator to specify the class name:

```
#include <iostream.h>
#include <string.h>
```

```
class employee {
 public:
   char name[64];
   long employee_id;
   float salary;
   void show_employee(void);
};

void employee::show_employee(void)
 {
   cout << "Name: " << name << endl;
   cout << "Id: " << employee_id << endl;
   cout << "Salary: " << salary << endl;
 };

void main(void)
 {
   employee worker, boss;

   strcpy(worker.name, "John Doe");
   worker.employee_id = 12345;
   worker.salary = 25000;

   strcpy(boss.name, "Happy Jamsa");
   boss.employee_id = 101;
   boss.salary = 101101.00;

   worker.show_employee();
   boss.show_employee();
 }
```

Understanding Class Methods

C++ classes let your programs group an object's data and the functions (methods) that operate on the data into one variable. When you define an object's methods, you have two choices. You can include all the function code within the class definition. Although including the object code in the method may seem convenient, as your classes become more complex and have several methods, the function statements may clutter the class definition. Therefore, many programs define the function statements outside of the class. Within the class definition, the program must include a function prototype that specifies the function name, return type, and parameter types.

To define a function outside of a class definition, your program must precede the function definition with the class name and the global resolution operator, as shown here:

```
return_type class_name::function_name(parameters)
 {
    // Statements
 }
```

LOOKING AT A SECOND EXAMPLE

The following program, *Pedigree.CPP*, creates a *dog* class, which contains several data fields and the *show_breed* function. The program defines the class function outside of the class definition itself. The program then creates three *dog* objects and displays information about each dog, as shown here:

```cpp
#include <iostream.h>
#include <string.h>

class dogs {
 public:
  char breed[64];
  int average_weight;
  int average_height;
  void show_breed(void);
};

void dogs::show_breed(void)
 {
   cout << "Breed: " << breed << endl;
   cout << "Average Weight: " << average_weight << endl;
   cout << "Average Height: " << average_height << endl;
 }

void main(void)
 {
   dogs happy, matt;

   strcpy(happy.breed, "Dalmatian");
   happy.average_weight = 58;
   happy.average_height = 24;

   strcpy(matt.breed, "Shetland Sheepdog");
   matt.average_weight = 22;
   matt.average_height = 15;

   happy.show_breed();
   matt.show_breed();
 }
```

WHAT YOU MUST KNOW

C++ programs make extensive use of classes. In short, a class lets your program group an object's data and the methods (function) that operate on that data into one variable. As you can see, classes are very much like the structures you examined in Lesson 20, "Storing Related Information in Structures." C++ classes provide the basis for object-oriented programming. The following lessons will expand on the different capabilities classes provide. As this lesson briefly discussed, the *public* label that appeared in the class definitions made the class members available throughout the entire program. In Lesson 24, you will learn more about *private* and *public* members within a class. Before you continue with Lesson 24, however, make sure that you have learned the following key concepts:

- ☑ In the simplest sense, an object is a thing upon which your program performs different operations.

- ☑ C++ programs represent objects using a class.

- ☑ A class, like a structure, contains members. Class members can store information (data) or be functions (methods) that operate on the data.

- ☑ Each class has a unique name.

- ☑ After you define a class, you can declare objects of that class using the class name as a type.

- ☑ To access class members (either data or functions), your programs use the dot operator.

- ☑ Your programs can define a class function within or outside of the class definition. If you define the function outside of the class definition, you must specify the class name and use a global resolution operator, such as *class::function*.

LESSON 24

UNDERSTANDING PUBLIC AND PRIVATE DATA

In Lesson 23, "Starting with C++ Classes," you created your first C++ classes. At that time, you included the *public* label within your class definition to provide your program with access to each of the class members. In this lesson you will learn how *public* and *private* class members control which class members your programs can access directly using the dot operator. As you will learn, your programs can access *public* members from within any function. On the other hand, your programs can only access *private* members using class functions. In this way, *private* class members let objects control how a program can use its data members. This lesson examines *public* and *private* members in detail. By the time you finish this lesson, you will understand the following key concepts:

- To control how your programs access class members, C++ lets you define members as *public* or *private*.

- *Private* members let a class hide information about that class which a program does not need to know or need to access directly.

- Classes that use *private* members provide *interface functions* that access the *private* members.

As briefly discussed in Lesson 23, when you define a class, you should place as much information about an object as you can within your class definition. In this way, your objects become self-contained, which can increase their reuse among multiple programs.

UNDERSTANDING INFORMATION HIDING

As you have learned, a class contains data and methods (functions). To use a class, your programs must know the information the class stores (its data members) and the methods that manipulate the data (the functions). Your programs do not need to know how the methods work. Rather, your programs must know only the task the methods perform. For example, assume that you have a *file* class.

Ideally, your program only must know that the class provides the methods *file.print*, which prints a formatted copy of the current file, and *file.delete*, which erases the file. Your program does not need to know how these two methods work. In other words, the program should treat the class as a "black box." The program knows which methods to invoke and the parameters to the methods, but the program does not know the actual processing that occurs within the class (the black box).

Information hiding is the process of making available to the program only the minimal class information the program needs in order to use a class. C++ *private* and *public* class members help you achieve information hiding within your programs. Each of the classes you created in Lesson 23 used the *public* label to make all the class members *public* or visible to the entire program. Therefore, the program could access any class member directly using the dot operator, as shown here:

```
class employee {
  public:
    char name[64];
    long employee_id;
    float salary;
    void show_employee(void);
}
```

159

When you create your classes, you might have members whose values the class uses internally to perform its processing, but that the program itself need not access. Such members are *private members* and you should hide them from the program. By default, if you do not include the *public* label, C++ assumes all class members are *private*. Your programs cannot access *private* class members using the dot operator. Only class-member functions can access *private* class members. When you create your classes, you will separate the members into *private* and *public* members, as shown here:

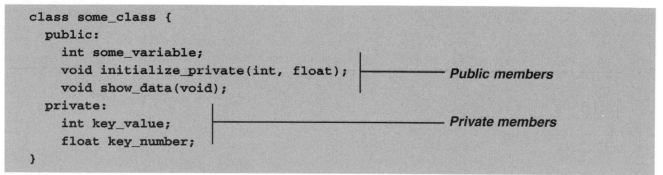

As you can see, the *public* and *private* labels let you define easily which members are *private* and which are *public*. In this example, the program can use the dot operator to access the *public* members, as shown here:

```
some_class object;        // Create an object
object.some_variable = 1001;
object.initialize_private(2002, 1.2345);
object.show_data()
```

If your program tries to access the *private* members *key_value* or *key_number* using the dot operator, the compiler will generate syntax errors.

As a general rule, you will usually protect class data members from direct program access by making the members *private*. In this way, programs cannot assign values to members directly using the dot operator. Instead, the program must call a class method to assign the values. By preventing the program's direct access to the data members, you can ensure that the program always assigns valid values to the class data members. For example, assume your program's *nuclear_reactor* object uses the member variable named *melt_down*, which should always contain a value in the range 1 through 5. If the *melt_down* member is *public*, the program can access the member's value directly, changing the value in any way it likes. For example, the following statement assigns the value 101 (which falls outside of the range 1 to 5) to the class member *melt_down*:

```
nuclear_reactor.melt_down = 101
```

If instead, you make the variable *private*, you can use a class method such as *assign_meltdown* to assign the member value. As shown here, the *assign_meltdown* function can test the value the program wants to assign to the member to ensure that the value is valid:

```
int nuke::assign_meltdown(int value)
  {
    if ((value > 0) && (value <= 5))
     {
       melt_down = value;
       return(0);   // Successful assignment
     }
    else
       return(-1);   // Invalid value
  }
```

Class methods that control access to class data members are *interface functions*. When you create classes you will use interface functions to protect your class data.

UNDERSTANDING PUBLIC AND PRIVATE MEMBERS

C++ classes contain data and methods. To control which members your programs can access directly using the dot operator, C++ lets you define *public* and *private* members. Your programs can directly access any *public* member using the dot operator. On the other hand, only class methods can access the *private* members. As a rule, you should protect most class data members by making them *private*. Then, the only way your programs can assign a value to a data member is to use a class function, which can examine and validate the value.

USING PUBLIC AND PRIVATE MEMBERS

The following program, *InfoHide.CPP*, illustrates the use of *public* and *private* members. The program defines an object of type *employee*, as shown here:

```cpp
class employee {
 public:
    int assign_values(char *, long, float);
    void show_employee(void);
    int change_salary(float);
    long get_id(void);
 private:
    char name[64];
    long employee_id;
    float salary;
}
```

As you can see, the class protects each of its data members by making them *private*. To access a data member, the program must use one of the *public* interface functions. The following program implements *InfoHide.CPP*:

```cpp
#include <iostream.h>
#include <string.h>

class employee {
 public:
    int assign_values(char *, long, float);
    void show_employee(void);
    int change_salary(float);
    long get_id(void);
 private:
    char name[64];
    long employee_id;
    float salary;
};

int employee::assign_values(char *emp_name,
  long emp_id, float emp_salary)
  {
```

```
    strcpy(name, emp_name);
    employee_id = emp_id;

    if (emp_salary < 50000.0)
     {
       salary = emp_salary;
       return(0);                    // Successful
     }
    else
      return(-1);                    // Invalid salary
 }

void employee::show_employee(void)
 {
   cout << "Employee: " << name << endl;
   cout << "Id: " << employee_id << endl;
   cout << "Salary: " << salary << endl;
 }

int employee::change_salary(float new_salary)
 {
   if (new_salary < 50000.0)
    {
      salary = new_salary;
      return(0);                     // Successful
    }
   else
     return(-1);                     // Invalid salary
 }

long employee::get_id(void)
 {
   return(employee_id);
 }

void main(void)
 {
   employee worker;

   if (worker.assign_values("Happy Jamsa", 101, 10101.0) == 0)
    {
      cout << "Employee values assigned" << endl;
      worker.show_employee();

      if (worker.change_salary(35000.00) == 0)
       {
         cout << "New salary assigned" << endl;
         worker.show_employee();
       }
    }
```

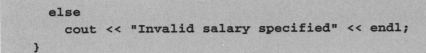

```
      else
         cout << "Invalid salary specified" << endl;
   }
```

Take time to examine the program statements in detail. Although the program is long, its functions are actually very straightforward. The *assign_values* method initializes the class *private* data. The method uses an *if* statement to ensure that it assigns a valid salary. The *show_employee* method displays the *private* data members. The *change_salary* and *get_id* methods are interface functions that provide the program with *private* data access. After you successfully compile and run the *InfoHide.CPP* program, edit the program and try to access directly a *private* data member using a dot operator from within *main*. Because you cannot access *private* members directly, the compiler will generate syntax errors.

UNDERSTANDING INTERFACE FUNCTIONS

To reduce potential errors, you should limit your program's access to class data by defining class-data members as *private*. In this way, a program cannot access class data members using the dot operator. Instead, the class should define interface functions with which the program can assign values to *private* members. The interface functions, in turn, can examine and validate the values that the program is trying to assign.

USING THE GLOBAL RESOLUTION OPERATOR FOR CLASS MEMBERS

If you examine the functions in the program *InfoHide.CPP*, you will find that the letters *emp_* often precede the function parameter names, as shown here:

```
int employee::assign_values(char *emp_name, long emp_id,
   float emp_salary
```

The functions use the letters *emp_* to prevent the parameter names from conflicting with the class member names. When such naming conflicts occur, you can resolve the conflicts by preceding the class member names with the class name and global resolution operator (::). The following function uses the global resolution operator and class name in front of the class member names. In this way, anyone reading the statements knows which names correspond to the *employee* class and which names refer to parameters, as shown here:

```
int employee::assign_values(char *name,
   long employee_id,  float salary)
 {
   strcpy(employee::name, name);
   employee::employee_id = employee_id;
   if (salary < 50000.0)
     {
       employee::salary = salary;
       return(0);                      // Successful
     }
   else
     return(-1);                       // Invalid salary
 }
```

As you create functions that work with class members, you should use the class name and global resolution operator in this way to avoid name conflicts.

Use the Global Resolution Operator to Resolve Class Members

When you create class member functions, there may be times when the local-variable name that you use within a function conflicts with a class-member name. By default, the local-variable name will override the class-member name. When such name conflicts occur, the function can use the class name and global resolution operator to access the class member, as shown here:

```
class_name::member_name = some_value;
```

PRIVATE MEMBERS ARE NOT ALWAYS DATA

In the examples this lesson presents, the *private* members were always data members. As your class definitions become more complex, you might have functions that other class methods use that you do not want the rest of your program to access directly. In such cases, you simply make the method a *private* member. If a class function is not *public*, the program cannot invoke the function using the dot operator. Instead, only other class members can invoke the *private* member functions.

WHAT YOU MUST KNOW

By controlling a program's access to class members, you reduce the possibility of errors that result from a program's member misuse. To control access to class members, you can use *private* members. Most of the C++ class definitions you will encounter will use a combination of *public* and *private* members. One of the most common operations your programs perform when they create an object is initializing the object data members. In Lesson 25, "Understanding Constructor and Destructor Functions," you will learn that C++ lets you define a special function called a *constructor* that C++ automatically invokes each time you create an object. Using the constructor function, your programs can initialize class members easily. Before you continue with Lesson 25, however, make sure that you have learned the following key concepts:

- ☑ Class members can be *public* or *private*. Programs can directly access *public* members using the dot operator. On the other hand, your programs can access private members using only class methods.

- ☑ Unless told otherwise, C++ assumes all members are *private*.

- ☑ Programs assign values to and access *private* class members using interface functions.

- ☑ When you create programs that manipulate class members, you can precede each member name with the class name and global resolution operator (::), such as *employee::name*, to resolve possible name conflicts.

LESSON 25

UNDERSTANDING CONSTRUCTOR AND DESTRUCTOR FUNCTIONS

When you create objects, one of the most common operations your programs will perform is to initialize the object's data members. As you learned in Lesson 24, "Understanding Private and Public Data," the only way your programs can access *private* data members is to use a class function. To simplify the process of initializing class-data members, C++ lets you define a special *constructor* function, on a per-class basis, that C++ automatically calls each time you create an object. In a similar way, C++ provides a *destructor* function that C++ runs when you discard an object. This lesson examines constructor and destructor functions in detail. By the time you finish this lesson, you will understand the following key concepts:

- Constructor functions are class methods that make it easy for your programs to initialize class-data members.

- Constructor functions have the same name as the class; yet you do not precede the constructor function name with the *void* keyword.

- Constructor functions do not return a type.

- Each time your program creates a class variable, C++ calls the class constructor function, if a constructor function exists.

- As your program runs, many objects may allocate memory to store information. When you discard an object, C++ will call a special destructor function that can free up such memory, in a sense, cleaning up after the object.

- Destructor functions have the same name as the class, with the exception that you must precede their name with the tilde character (~).

- Destructor functions do not return a type. Like a constructor function, you do not precede the destructor function name with the *void* keyword.

Do not let the terms constructor and destructor intimidate you. Instead, think of a constructor function as a function that helps you build (construct) an object. Likewise, a destructor is a function that helps you destroy an object. The use of destructor functions is most common when an object allocates memory you want the object to release before your program destroys the object.

CREATING A SIMPLE CONSTRUCTOR FUNCTION

A *constructor function* is a class method that has the same name as the class itself. For example, if you are using a class named *employee*, the constructor function's name is also *employee*. Likewise, for a class named *dogs*, the constructor function's name is *dogs*. If your program defines a constructor function, C++ will automatically invoke the function each time you create an object of the corresponding class type. The following program, *ConStruc.CPP*, creates a class named *employee*. The program also defines a constructor function named *employee* that assigns the object's initial values. A constructor function cannot return a value; however, you do not declare the function as *void*. Instead, you do not specify a return type, as shown here:

```
class employee {
  public:
```

165

```
  employee(char *, long, float); // Constructor function (no return type)
  void show_employee(void);
  int change_salary(float);
  long get_id(void);
 private:
  char name[64];
  long employee_id;
  float salary;
};
```

Within your program, you define the constructor function just as you would any class method, as shown here:

```
employee::employee(char *name, long employee_id, float salary)
 {
   strcpy(employee::name, name);
   employee::employee_id = employee_id;
   if (salary < 50000.0)
     employee::salary = salary;
   else
     // Invalid salary specified
     employee::salary = 0.0;
 }
```

As you can see, the constructor function does not return a value to the caller. Also, the function does not use the type *void*. In this example, the function uses the global resolution operator and class name before each member. The following program implements the complete *ConStruc.CPP*:

```
#include <iostream.h>
#include <string.h>

class employee {
 public:
   employee(char *, long, float);
   void show_employee(void);
   int change_salary(float);
   long get_id(void);
 private:
   char name[64];
   long employee_id;
   float salary;
};

employee::employee(char *name, long employee_id, float salary)
 {
   strcpy(employee::name, name);
   employee::employee_id = employee_id;
   if (salary < 50000.0)
     employee::salary = salary;
   else
     // Invalid salary specified
```

```
        employee::salary = 0.0;
   }


void employee::show_employee(void)
   {
      cout << "Employee: " << name << endl;
      cout << "Id: " << employee_id << endl;
      cout << "Salary: " << salary << endl;
   }


void main(void)
   {
      employee worker("Happy Jamsa", 101, 10101.0);


      worker.show_employee();
   }
```

Note that the *ConStruc.CPP* program follows declaration of the object *worker* with parentheses and the object's initial values, just like a function call. When your program uses constructor functions, your program can pass parameters to a constructor when it declares an object, as shown here:

```
   employee worker("Happy Jamsa", 101, 10101.0);
```

If your program had created several *employee* objects, you could initialize each object's members using the constructor, as shown here:

```
      employee worker("Happy Jamsa", 101, 10101.0);
      employee secretary("John Doe", 57, 20000.0);
      employee manager("Jane Doe", 1022, 30000.0:
```

Understanding Constructor Functions

 A constructor function is a special function C++ automatically invokes each time you create an object. The most common use of a constructor is to initialize the object's data members. Constructor functions have the same name as the object class. A class named *file*, for example, uses a constructor named *file*. You define constructor functions within your program just as you would any class method. The only difference is that for constructor functions you do not specify a return type. When you later declare an object, you can pass parameters to the constructor, as shown here:

```
   class_name  object(value1, value2, value3)
```

SPECIFYING DEFAULT PARAMETER VALUES FOR CONSTRUCTOR FUNCTIONS

As you learned in Lesson 17, "Providing Default Parameter Values," C++ lets you specify default function parameter values. If the user does not specify values for each parameter, the function will use the default values. Constructor functions are no exception; your programs can specify default values just as they would for any function. For example, the following *employee* constructor function uses the default salary of 10000.00 if the program does not specify a salary when it creates the object. The program must, however, specify an employee name and identification number:

```
employee::employee(char *name, long employee_id, float salary = 10000.00)
  {
    strcpy(employee::name, name);
    employee::employee_id = employee_id;
    if (salary < 50000.0)
      employee::salary = salary;
    else
      // Invalid salary specified
      employee::salary = 0.0;
  }
```

OVERLOADING CONSTRUCTOR FUNCTIONS

As you learned in Lesson 15, "Overloading Functions," C++ lets your programs overload function definitions by specifying alternative functions for different parameter types. C++ also lets you overload constructor functions. The following program, *ConsOver.CPP*, overloads the *employee* constructor function. The first constructor function definition requires the program to specify an employee name, identification number, and salary. The second constructor function definition prompts the user to type in the desired salary if the program does not specify one, as shown here:

```
employee::employee(char *name, long employee_id)
  {
    strcpy(employee::name, name);
    employee::employee_id = employee_id;
    do {
      cout << ""Enter a salary for " << name <<
        " less than $50,000: ";
      cin >> employee::salary;
    } while (salary >= 50000.0);
  }
```

Within the class definition, the program must specify both function prototypes as shown here:

```
class employee {
 public:
    employee(char *, long, float);
    employee(char *, long);          |———— Overload function prototypes
    void show_employee(void);
    int change_salary(float);
    long get_id(void);
 private:
    char name[64];
    long employee_id;
    float salary;
}
```

The following statements implement the *ConsOver.CPP* program:

```
#include <iostream.h>
#include <string.h>
```

```cpp
class employee {
public:
  employee(char *, long, float);
  employee(char *, long);
  void show_employee(void);
  int change_salary(float);
  long get_id(void);
private:
  char name[64];
  long employee_id;
  float salary;
};

employee::employee(char *name, long employee_id,
  float salary)
 {
   strcpy(employee::name, name);
   employee::employee_id = employee_id;
   if (salary < 50000.0)
     employee::salary = salary;
   else
     // Invalid salary specified
     employee::salary = 0.0;
 }

employee::employee(char *name, long employee_id)
 {
   strcpy(employee::name, name);
   employee::employee_id = employee_id;

   do {
     cout << "Enter a salary for " << name <<
       " less than $50,000: ";
     cin >> employee::salary;
   } while (salary >= 50000.0);
 }

void employee::show_employee(void)
 {
   cout << "Employee: " << name << endl;
   cout << "Id: " << employee_id << endl;
   cout << "Salary: " << salary << endl;
 }

void main(void)
 {
   employee worker("Happy Jamsa", 101, 10101.0);
   employee manager("Jane Doe", 102);
```

```
      worker.show_employee();
      manager.show_employee();
   }
```

When you compile and execute the *ConsOver.CPP* program, your screen will prompt you to type in a salary amount for Jane Doe. When you type in the amount, the program's execution will continue displaying information about both employees.

UNDERSTANDING DESTRUCTOR FUNCTIONS

Just as C++ lets you define a constructor function, which C++ automatically calls when you create a class object, C++ also lets you define a *destructor function,* which it calls when you destroy the object. In later lessons, you will learn how to create lists of objects that grow and shrink as your program executes. To create such dynamic lists, for example, your program will allocate memory dynamically to hold the objects (something you have not yet learned to do). At that time, you may create and destroy objects throughout your program's execution. Then, using destructor functions should make more sense.

Each of the programs you have created so far create objects when the program begins, by declaring the objects. When your programs end, C++ destroys the objects. If you define a destructor function within your functions, C++ will automatically invoke the destructor for each object as the program ends (as C++ destroys the objects). Like constructor functions, destructor functions have the same name as the object class. However, for destructor functions, you precede the name with the tilde character (~), as shown here:

```
~class_name(void)          ~ indicates destructor
  {
     // Function statements
  }
```

Unlike constructor functions, you cannot pass parameters to a destructor function.

The following program, *Destruct.CPP*, defines the *employee* class destructor function as follows:

```
void employee::~employee(void)
  {
    cout << "Destroying the object for " << name << endl;
  }
```

In this example, the destructor function simply displays a message on your screen that C++ is destroying the object. When the program ends, C++ automatically invokes the function for each object. The following program implements *Destruct.CPP*:

```
#include <iostream.h>
#include <string.h>

class employee {
 public:
    employee(char *, long, float);
    ~employee(void);
    void show_employee(void);
    int change_salary(float);
```

```
   long get_id(void);
 private:
   char name[64];
   long employee_id;
   float salary;
};

employee::employee(char *name, long employee_id,
  float salary)
 {
   strcpy(employee::name, name);
   employee::employee_id = employee_id;
   if (salary < 50000.0)
     employee::salary = salary;
   else
     // Invalid salary specified
     employee::salary = 0.0;
 }


 employee::~employee(void)
 {
   cout << "Destroying the object for " << name << endl;
 }

void employee::show_employee(void)
 {
   cout << "Employee: " << name << endl;
   cout << "Id: " << employee_id << endl;
   cout << "Salary: " << salary << endl;
 }

void main(void)
 {
   employee worker("Happy Jamsa", 101, 10101.0);

   worker.show_employee();
 }
```

When you compile and execute the *Destruct.CPP* program, your screen will display the following output:

```
C:\> Destruct   <ENTER>
Employee: Happy Jamsa
Id: 101
Salary: 10101
Destroying the object for Happy Jamsa
```

As you can see, the program automatically invokes the destructor function without using a function call. For now, your programs probably will not use destructor functions. When your programs begin to allocate memory within objects, however, you may find that destructor functions provide a convenient way to release the memory when your programs destroy an object.

171

Understanding Destructor Functions

A destructor function is a function that C++ automatically executes when it or your program destroys an object. Destructor functions have the same name as the object class; however, you precede the destructor function name with a tilde (~) character, as in *~employee*. Within your program, you define destructor methods just as you would any class method.

WHAT YOU MUST KNOW

Constructors and destructors are special class functions C++ automatically invokes when your program creates or destroys an object. Most programs use constructor functions to initialize class data members. The simple programs you are currently creating probably will not require destructor functions. In Lesson 26, "Understanding Operator Overloading," you will learn how to overload operators. In other words, you may redefine the plus (+) symbol so that it can append the contents of one string to another. As you have learned, a type (such as *char*, *float*, and *int*) defines a set of values a variable can store and a set of operations your programs can perform on the variable. When you define a class, you essentially define a type. C++ lets you specify how operators behave for that type. Before you continue with Lesson 26, however, make sure that you have learned the following key concepts:

☑ A constructor function is a special function C++ automatically invokes each time your program creates an object. Constructor functions have the same name as the object class.

☑ Constructor functions do not return a value, but you do not define them as type *void*. Instead, you simply do not specify a return value type.

☑ When your programs create an object, your programs can pass parameters to the constructor function during the object's declaration.

☑ C++ lets you overload constructor functions and provide default parameter values.

☑ A destructor function is a special function that C++ automatically invokes each time your program creates or destroys an object. Destructor functions also have the same name as the object class, but the function name is preceded by a tilde (~) character.

LESSON 26

UNDERSTANDING OPERATOR OVERLOADING

As you have learned, a variable's type specifies the set of values a variable can store and a set of operations you can perform on the variable. For example, using a variable of type *int*, your program can add, subtract, multiply, and divide values. With a character string, on the other hand, using the plus operator to add two strings would not make sense. When you define a class within your programs, you essentially define a new type. Therefore, C++ lets you specify your new type's corresponding operations. *Operator overloading* is the process of changing the meaning of an operator (such as the plus operator (+), which C++ normally uses for addition) for use by a specific class. In this lesson you will define a *string* class and overload the plus and minus operators. For string objects, the plus operator will append specified characters to the string's current contents. In a similar way, the minus operator will remove each occurrence of a specified letter from the string. By the time you finish this lesson, you will understand the following key concepts:

- You overload operators to improve your program's readability. You should overload an operator, however, only when doing so makes your program easier to understand.

- Your programs use the C++ *operator* keyword to overload an operator.

- When you overload an operator, you specify a function that C++ invokes each time the class uses the overloaded operator. The function, in turn, performs the corresponding operation.

- When your programs overload an operator for a specific class, the operator's meaning changes for that class only—the rest of your program will continue to use the operator to perform its standard operations.

- C++ will let your programs overload most operators; however, there are four operators (see Table 26) that your programs cannot overload.

Operator overloading can simplify common class operations and improve your program's readability. Take time to experiment with the programs this lesson presents and you will find that operator overloading is quite easy.

OVERLOADING THE PLUS AND MINUS OPERATORS

When you overload an operator for a class, that operator's function does not change for other variable types. For example, if you overload the plus operator for the *string* class, the operator's function will not change when you must add two numbers. When the C++ compiler encounters the operator within your program, the compiler will determine which operation to perform based on the corresponding variable's type. The following class definition creates a *string* class. The class contains one data member, which is the character string itself. The class contains several different methods and does not currently define any operators, as shown here:

```
class string {
  public:
    string(char *);  // Constructor
    void str_append(char *);
    void chr_minus(char);
    void show_string(void);
  private:
    char data[256];
};
```

As you can see, the class definition provides the *str_append* function, which appends specified characters to the contents of the class string. Likewise, the function *chr_minus* removes each occurrence of a specified character from the class string. The following program, *StrClass.CPP*, uses the *string* class to create and manipulate two character-string objects:

```
#include <iostream.h>
#include <string.h>

class string {
 public:
    string(char *);   // Constructor
    void str_append(char *);
    void chr_minus(char);
    void show_string(void);
 private:
    char data[256];
};

string::string(char *str)
  {
    strcpy(data, str);
  }

void string::str_append(char *str)
  {
    strcat(data, str);
  }

void string::chr_minus(char letter)
  {
    char temp[256];
    int i, j;

    for (i = 0, j = 0; data[i]; i++)
      // Is the letter to remove?
      if (data[i] != letter)
        // If not assign it to temp
        temp[j++] = data[i];

    temp[j] = NULL;   // End of temp

    // Copy temp's contents back to data
    strcpy(data, temp);
  }

void string::show_string(void)
  {
    cout << data << endl;
  }
```

```
void main(void)
{
    string title("Rescued By C++");
    string lesson("Understanding Operator Overloading");

    title.show_string();
    title.str_append(" rescued me!");
    title.show_string();

    lesson.show_string();
    lesson.chr_minus('n');
    lesson.show_string();
}
```

As you can see, the program uses the function *str_append* to append characters to the string variable *title*. Also, the program uses the function *chr_minus* to remove each occurrence of the letter *n* from the *lesson* character string. In this example, the *StrClass.CPP* program uses function calls to perform these operations. Using operator overloading, however, the program can perform identical operations using the plus (+) and minus (-) operators.

When you overload an operator, you use the C++ *operator* keyword within the function prototype and definition to tell the C++ compiler that the class will use this method as an operator. The following class definition, for example, uses the *operator* keyword to assign the plus and minus operators to the functions *str_append* and *chr_minus* within the *string* class:

```
class string {
 public:
    string(char *);   // Constructor
    void operator +(char *);  ─┐
                               ├── Defining class operators
    void operator -(char);    ─┘
    void show_string(void);
 private:
    char data[256];
};
```

As you can see, the class overloads the plus and minus operators. As briefly discussed, when a class overloads an operator, the class must specify a function that implements the operator's corresponding operation. In the case of the plus operator, the function definition becomes the following:

```
void string::operator +(char *str)
{
    strcat(data, str);
}
```

As you can see, the function definition does not specify a name, but rather, the operator class as overloading. To overload the plus operator, the program has not changed the processing that occurs within the function (this function's code is identical to the code in the previous *str_append* function). Instead, the program simply replaced the function name with the *operator* keyword and corresponding operator. The following program, *OpOverld.CPP*, illustrates the use of the overloaded plus and minus operators:

```
#include <iostream.h>
#include <string.h>
```

```cpp
class string {
 public:
   string(char *);   // Constructor
   void operator +(char *);
   void operator -(char);
   void show_string(void);
 private:
   char data[256];
};

string::string(char *str)
 {
   strcpy(data, str);
 }

void string::operator +(char *str)
 {
   strcat(data, str);
 }

void string::operator -(char letter)
 {
   char temp[256];
   int i, j;

   for (i = 0, j = 0; data[i]; i++)
     if (data[i] != letter)
       temp[j++] = data[i];

   temp[j] = NULL;

   strcpy(data, temp);
 }

void string::show_string(void)
 {
   cout << data << endl;
 }

void main(void)
 {
   string title("Rescued By C++");
   string lesson("Understanding Operator Overloading");

   title.show_string();
   title + " rescued me!";
   title.show_string();

   lesson.show_string();
```

```
    lesson - 'n';
    lesson.show_string();
  }
```

You can see that the *OpOverld.CPP* program uses the overloaded operators, as shown here:

```
// Append the text "rescued me!"
title + " rescued me!";
lesson - 'n';                              // Remove the letter 'n'
```

In this example, the operator syntax is legal, but a little strange. Normally programs use the plus operator in an expression that returns a result, such as the *some_str = title + "text";* statement. When you define your operators, C++ gives you quite a bit of freedom in determining how the operator behaves. However, as you will recall, your purpose in overloading operators is to make your programs easy to understand. For example, the following program, *Str_Over.CPP*, changes the previous program slightly to let your programs perform operations on the variables of type *string* using a syntax that is more consistent with standard assignment operations:

```
#include <iostream.h>
#include <string.h>

class string {
 public:
    string(char *);   // Constructor
    char * operator +(char *);
    char * operator -(char);
    void show_string(void);
 private:
    char data[256];
};

string::string(char *str)
  {
    strcpy(data, str);
  }

char * string::operator +(char *str)
  {
    return(strcat(data, str));
  }

char * string::operator -(char letter)
  {
    char temp[256];
    int i, j;

    for (i = 0, j = 0; data[i]; i++)
      if (data[i] != letter)
        temp[j++] = data[i];
    temp[j] = NULL;
```

177

```
        return(strcpy(data, temp));
    }

void string::show_string(void)
    {
      cout << data << endl;
    }

void main(void)
    {
      string title("Rescued By C++");
      string lesson("Understanding Operator Overloading");

      title.show_string();
      title = title + " rescued me!";
      title.show_string();

      lesson.show_string();
      lesson = lesson - 'n';
      lesson.show_string();
    }
```

By changing the overloaded plus and minus operators to return pointers to character strings, the program can now use the overloaded operators using an assignment-statement syntax, as shown here:

```
      title = title + " rescued me!";
      lesson = lesson - 'n';
```

If you examine the code closely, you will find the previous statements assign a pointer to a character array to the *string* class. When the C++ compiler encounters the assignment, C++ invokes the *string* class constructor function which assigns character array's contents to the object's *data* member.

LOOKING AT ANOTHER EXAMPLE

When you create your own data types using classes, a common operation you will perform is testing to determine if two objects are the same. Using operator overloading, your programs can overload the equal (==) , not equal (!=), or other relational operators. The following program, *Comp_Str.CPP*, adds a new operator to the *string* class that tests whether two *string* objects are equal. Using the overloaded operator, your programs can test whether string objects contain the same string as follows:

```
    if (some_string == another_string)
```

The following statements implement the *Comp_Str.CPP* program:

```
    #include <iostream.h>
    #include <string.h>

    class string {
     public:
       string(char *);   // Constructor
```

```
    char * operator +(char *);
    char * operator -(char);
    int operator ==(string);
    void show_string(void);
 private:
    char data[256];
};

string::string(char *str)
  {
    strcpy(data, str);
  }

char * string::operator +(char *str)
  {
    return(strcat(data, str));
  }

char * string::operator -(char letter)
  {
    char temp[256];
    int i, j;

    for (i = 0, j = 0; data[i]; i++)
      if (data[i] != letter)
        temp[j++] = data[i];

    temp[j] = NULL;

    return(strcpy(data, temp));
  }

int string::operator ==(string str)
  {
    int i;

    for (i = 0; data[i] == str.data[i]; i++)
      if ((data[i] == NULL) && (str.data[i] == NULL))
        return(1);   // Equal

    return(0);       // Not equal
  }

void string::show_string(void)
  {
    cout << data << endl;
  }

void main(void)
  {
```

```
string title("Rescued By C++");
string lesson("Understanding Operator Overloading");
string str("Rescued By C++");

if (title == lesson)
  cout << "title and lesson are equal" << endl;


if (str == lesson)
  cout << "str and lesson are equal" << endl;


if (title == str)
  cout << "title and str are equal" << endl;
}
```

As you can see, by overloading operators in this way, you make your programs easier to understand.

OPERATORS YOU CANNOT OVERLOAD

In general, your programs can overload almost every C++ operator. Table 26 lists the operators C++ will not let your programs overload.

Operator	Purpose	Example
.	Class member operator	object.member
.*	Pointer to member operator	object.*member
::	Global scope resolution operator	classname::member
?:	Conditional expression operator	c = (a > b) ? a : b;

Table 26 C++ operators your programs cannot overload.

WHAT YOU MUST KNOW

Operator overloading is the ability to assign a new meaning to an operator when a specific class uses the operator. Using operator overloading, you can improve your program's readability and ease of understanding by expressing class operations in a more meaningful way. In Lesson 27, "Static Function and Data Members," you will learn how to share data between objects using a *static* member and how to use a class method when the program has not declared any class objects. Before you continue with Lesson 27, however, make sure that you have learned the following key concepts:

☑ To overload an operator, you must define the class to which you will assign the operator.

☑ When you overload an operator, the overload is in effect only for a specific class. If the program uses the operator with nonclass variables (such as variables of type *int* or *float*), C++ uses the operator's original definition.

☑ To overload a class operator, you use the C++ *operator* keyword to define the class method C++ invokes each time a class variable uses the operator.

☑ C++ does not let your programs overload the member operator (.), pointer to member operator (.*), the scope resolution operator (::), or the conditional expression operator (?:).

LESSON 27

STATIC FUNCTION AND DATA MEMBERS

Up to this point, each object your programs have created has received its own private set of data members. Depending on your program's purpose, there may be times when you will want objects of the same class to share one or more data variables. For example, assume you are writing a payroll program that tracks the hours for 1,000 workers. To determine tax rates, the program must know the state in which each employee works. Therefore, one of the class members is *state_of_work*. However, if all the employees work in the same state, your program could share the data member among all employee objects.

In this way, your program reduces the amount of memory it requires by eliminating 999 copies of the same information. To share a class member, you declare the member as *static*. This lesson examines the steps you must perform to share class members among objects. By the time you finish this lesson, you will understand the following key concepts:

- C++ lets objects of the same class type share one or more class members.

- When your program assigns a value to a shared member, all objects of that class type immediately have access to the new value.

- To create a shared class data member, you precede the member name with the *static* keyword.

- After your program declares a class member as *static*, your program must declare a global variable (outside of the class definition) that corresponds to the shared class member.

- Your programs can use the *static* keyword to make a class method callable, although the program may not yet have defined any objects of the class type.

SHARING A DATA MEMBER

Normally, when you create objects of a specific class type, each object gets its own set of data members. However, there may be times when objects of the same class must share one or more data members. In such cases, declare the data members as *public* or *private* members and precede the type with the *static* keyword, as shown here:

```
private:
  static int shared_value;
```

After you declare the class, you must define the member as a global variable, outside of the class, as shown here:

```
int class_name::shared_value;
```

The following program, *Share_It.CPP*, defines the *book_series* class, which shares the *page_count* member, which is the same for all objects (books) of the class (series). If the program changes the member's value, all objects of the class will immediately see the change, as shown here:

```
#include <iostream.h>
#include <string.h>

class book_series {
  public:
```

181

```
    book_series(char *, char *, float);
    void show_book(void);
    void set_pages(int);
  private:
    static int page_count;
    char title[64];
    char author[64];
    float price;
};

int book_series::page_count;

void book_series::set_pages(int pages)
 {
   page_count = pages;
 }

book_series::book_series(char *title,
  char *author, float price)
 {
   strcpy(book_series::title, title);
   strcpy(book_series::author, author);
   book_series::price = price;
 }

void book_series::show_book(void)
 {
   cout << "Title: " << title << endl;
   cout << "Author: " << author << endl;
   cout << "Price: " << price << endl;
   cout << "Pages: " << page_count << endl;
 }

void main(void)
 {
   book_series programming("Rescued by C++, Third Edition", "Jamsa",
     29.95);
   book_series upgrade("Rescued by Upgrading Your PC, Second Edition",
     "Jamsa", 24.95);

   upgrade.set_pages(256);

   programming.show_book();
   upgrade.show_book();

   cout << endl << "Changing page count " << endl;

   programming.set_pages(280);

   programming.show_book();
   upgrade.show_book();
 }
```

As you can see, the class declares the member *page_count* as *static int*. Immediately following the class definition, the program declares the *page_count* member as a global variable. When the program changes the *page_count* member, all objects of the *book_series* class immediately see the change.

Sharing Class Members

Depending on your program, there may be times when you must share specific data among object instances. To do so, declare the member as *static*. Next, declare the member outside of the class as a global variable. Later, all objects of the class type will immediately see any changes your program makes to the member.

USING PUBLIC STATIC MEMBERS WHEN NO OBJECTS EXIST

As you just learned, when you declare a class member as *static* all objects of the class type share the member. There may be times, however, when your programs have not yet created an object, but the program needs to use the member. To use the member, your program must declare the member as *public* and *static*. For example, the following program, *Use_Mbr.CPP*, uses the *page_count* member from the *book_series* class, even though no objects of the class exist:

```cpp
#include <iostream.h>
#include <string.h>

class book_series {
 public:
   static int page_count;
 private:
   char title[64];
   char author[64];
   float price;
};

int book_series::page_count;

void main(void)
  {
   book_series::page_count = 256;

   cout << "The current page count is " <<
     book_series::page_count << endl;
  }
```

In this example, because the class defines the class member *page_count* as *public*, the program can access the class member even though no objects of the *book_series* class exist.

USING STATIC MEMBER FUNCTIONS

The previous programs have illustrated the use of *static* data members. In a similar way, C++ lets you define the member functions (methods) as *static*. When you create a *static* method, your program can call the function even if no objects exist. If, for example, a class contains a method that the program may use to manipulate data outside of **183** the class, you might make the method *static*. For example, the following *menu* class uses an ANSI driver escape sequence to clear the screen display. If your system has the *ANSI.SYS* device driver installed, you can use the *clear_screen* method to clear the screen. Because you declare the method as *static* the program can use the method, even though no objects of type *menu* exist. The following program, *Clr_Scr.CPP*, uses the *clear_screen* method to clear the screen display:

```cpp
#include <iostream.h>

class menu {
 public:
   static void clear_screen(void);
   // Other methods would be here
 private:
   int number_of_menu_options;
};

void menu::clear_screen(void)
 {
   cout << '\033' << "[2J";
 }

void main(void)
 {
   menu::clear_screen();
 }
```

Because the program declares the *clear_screen* member as *static*, the program can use the function to clear the screen, even though no objects of type *menu* exist. The *clear_screen* function uses the ANSI escape sequence *Esc[2J* to clear the screen. For more information on using ANSI escape sequences in your programs, refer to the book *Jamsa's C/C++ Programmer's Bible*, Jamsa Press, 1997.

Letting Your Programs Use Class Methods

As you create class methods, there may be times when a function that you create for use by a class may be useful for operations within your program that do not involve class objects. For example, the previous *menu* class defined a *clear_screen* function that you may want to use throughout your program. When your class contains a method that you may want to use outside of a class object, precede the method prototype with the *static* keyword and declare the method as *public,* as shown here:

```cpp
public:
  static void clear_screen(void);
```

Within your program, you use the global resolution operator to invoke the function, as shown here:

```cpp
menu::clear_screen();
```

WHAT YOU MUST KNOW

In this lesson you learned that if you precede a class data member with the *static* keyword you make the member shared among all objects of the class type. If the data member is *public*, your programs can access the data member's value even though no objects of the class type may exist. Likewise, if your programs precede a *public* class method with the *static* keyword, your programs can use the function for operations that do not involve class objects. In Lesson 28, "Understanding Inheritance," you will learn how to use inheritance to build one object from one or more existing objects. Using inheritance to create new objects can save you considerable programming. Before you continue with Lesson 28, however, make sure that you have learned the following key concepts:

☑ When you declare a class member as *static*, the member becomes shared among all objects of the class type.

☑ After your program declares a class member as *static*, the program must declare a global variable outside of the class definition that corresponds to the shared class member.

☑ If you declare a member as *public* and *static*, your program can use the member, even though no objects of the class exist. To access the member, your program must use a global resolution operator, such as *class_name::member_name*.

☑ If you declare a *public static* function member, your program can invoke the function, even though no objects of the class exist. To call the function, your programs must use a global resolution operator, such as *menu::clear_screen()*.

STANDARD TEMPLATE LIBRARY

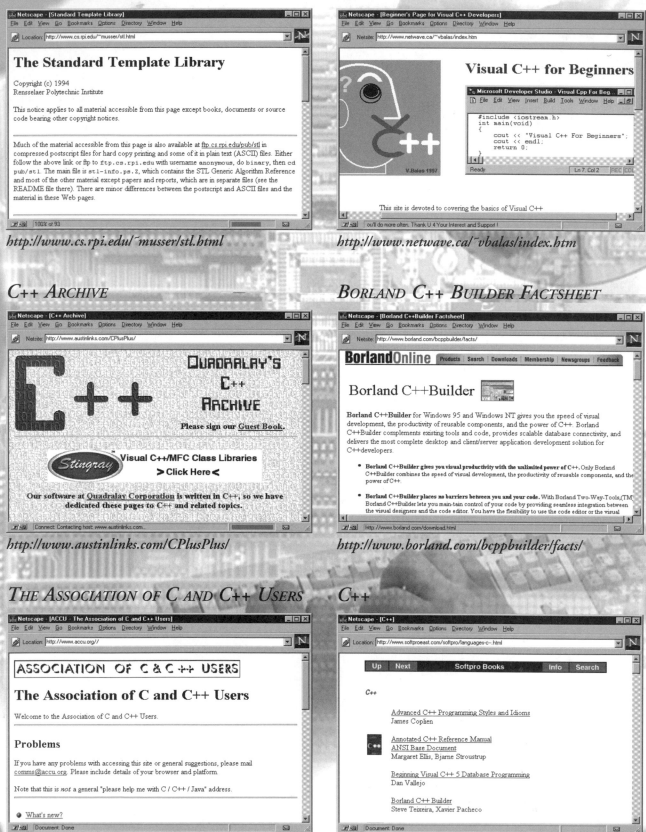

The Standard Template Library

Copyright (c) 1994
Rensselaer Polytechnic Institute

This notice applies to all material accessible from this page except books, documents or source code bearing other copyright notices.

Much of the material accessible from this page is also available at ftp.cs.rpi.edu/pub/stl in compressed postscript files for hard copy printing and some of it in plain text (ASCII) files. Either follow the above link or ftp to ftp.cs.rpi.edu with username anonymous, do binary, then cd pub/stl. The main file is stl-info.ps.z, which contains the STL Generic Algorithm Reference and most of the other material except papers and reports, which are in separate files (see the README file there). There are minor differences between the postscript and ASCII files and the material in these Web pages.

http://www.cs.rpi.edu/~musser/stl.html

BEGINNER'S PAGE FOR VISUAL C++ DEVELOPERS

Visual C++ for Beginners

```
#include <iostream.h>
int main(void)
{
    cout << "Visual C++ For Beginners";
    cout << endl;
    return 0;
}
```

V.Balas 1997

This site is devoted to covering the basics of Visual C++

http://www.netwave.ca/~vbalas/index.htm

C++ ARCHIVE

QUADRALAY'S C++ ARCHIVE

Please sign our Guest Book.

Stingray™ Visual C++/MFC Class Libraries
> Click Here <

Our software at Quadralay Corporation is written in C++, so we have dedicated these pages to C++ and related topics.

http://www.austinlinks.com/CPlusPlus/

BORLAND C++ BUILDER FACTSHEET

BorlandOnline Products | Search | Downloads | Membership | Newsgroups | Feedback

Borland C++Builder

Borland C++Builder for Windows 95 and Windows NT gives you the speed of visual development, the productivity of reusable components, and the power of C++. Borland C++Builder complements existing tools and code, provides scalable database connectivity, and delivers the most complete desktop and client/server application development solution for C++developers.

- **Borland C++Builder gives you visual productivity with the unlimited power of C++.** Only Borland C++Builder combines the speed of visual development, the productivity of reusable components, and the power of C++.

- **Borland C++Builder places no barriers between you and your code.** With Borland Two-Way-Tools,(TM) Borland C++Builder lets you main-tain control of your code by providing seamless integration between the visual designers and the code editor. You have the flexibility to use the code editor or the visual

http://www.borland.com/bcppbuilder/facts/

THE ASSOCIATION OF C AND C++ USERS

ASSOCIATION OF C & C++ USERS

The Association of C and C++ Users

Welcome to the Association of C and C++ Users.

Problems

If you have any problems with accessing this site or general suggestions, please mail comms@accu.org. Please include details of your browser and platform.

Note that this is *not* a general "please help me with C / C++ / Java" address.

● What's new?

http://www.accu.org//

C++

| Up | Next | Softpro Books | Info | Search |

C++

Advanced C++ Programming Styles and Idioms
James Coplien

Annotated C++ Reference Manual
ANSI Base Document
Margaret Ellis, Bjarne Stroustrup

Beginning Visual C++ 5 Database Programming
Dan Vallejo

Borland C++ Builder
Steve Teixeira, Xavier Pacheco

http://www.softproeast.com/softpro/languages-c--.html

SECTION FIVE

INHERITANCE AND TEMPLATES

A major advantage of object-oriented programming is that you can often use a class you create for one program in another. As you will learn in this section, C++ not only lets your programs reuse classes, it lets you build one class from another. When you build one class from another, the new class will inherit the characteristics of the original class. In this section you will learn how to use C++ inheritance capabilities to save considerable programming. In addition, you will learn how C++ templates make it easy for you to define generic data types. For example, using a template, you can define a generic *array* class that you later use to create one variable that can store values of type *int* and a second variable that can store values of type *float*. By the time you finish this section, you will have learned a great deal about object-oriented programming concepts. The lessons in this section include:

LESSON 28

UNDERSTANDING INHERITANCE

Object-oriented programming makes it easy for you to reuse a class that you create for one program within another program, thereby saving you time and programming. As you define your classes, there may be times when a class uses many or all the features of an existing class and then adds one or more members, either data or function members. In such cases, C++ lets you build the new object using the characteristics of the existing object. In other words, the new object will *inherit* the members of the existing class (called the *base class*). When you build a new class from an existing class, the new class is a *derived class*. This lesson introduces C++ class inheritance. By the time you finish this lesson, you will understand the following key concepts:

- When your programs use inheritance, they use a *base class* to derive a new class. The new (derived) class inherits the base-class members.

- To initialize derived class members, your program must invoke the base-class and derived-class constructor functions.

- Using the dot operator, your programs can easily access base-class and derived-class members.

- In addition to *public* (accessible to all) and *private* (accessible to class methods) members, C++ provides *protected* members that the base- and derived-class members can access.

- To resolve name conflicts between base- and derived-class members, your program can use the global-resolution operator (::), preceding the operator with the base- or derived-class name.

Inheritance is a fundamental object-oriented programming concept. Take time to experiment with the programs this lesson presents. As you will learn, inheritance is actually very easy to implement and can save you many hours of programming.

A SIMPLE INHERITANCE

Inheritance lets a derived class inherit the characteristics of an existing base class. Assume, for example, that you have the following *employee* base class:

```
class employee {
public:
    employee(char *, char *, float);
    void show_employee(void);
private:
    char name[64];
    char position[64];
    float salary;
};
```

Next, assume that your program needs a *manager* class, which adds the following data members to the *employee* class:

```
float annual_bonus;
char company_car[64];
int stock_options;
```

In this example, your program has two choices. First, the program can create a new *manager* class, which duplicates many of the *employee* class members. Second, your program can derive the class type *manager* from the *employee* base class. By deriving the *manager* class from the existing *employee* class, you reduce your programming and eliminate duplicate code within your program. To start the class definition, you will specify the *class* keyword, the *manager* name followed by a colon, and the *employee* name, as shown here:

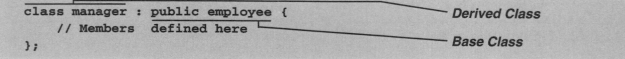

```
class manager : public employee {
    // Members  defined here
};
```

Derived Class

Base Class

The *public* keyword, which precedes the *employee* class name, specifies that the *public* members in the *employee* class are *public* within the *manager* class as well. For example, the following statements derive the *manager* class:

```
class manager : public employee {
 public:
    manager(char *, char *, char *, float, float, int);
    void show_manager(void);
 private:
    float annual_bonus;
    char company_car[64];
    int stock_options;
};
```

When you derive a class from a base class, the *private* members within the base class are accessible only to the derived class through the interface functions within the base class. Therefore, the derived class cannot access a base-class *private* member directly using the dot operator. The following program, *Mgr_Emp.CPP*, illustrates how to use C++ inheritance, building the *manager* class from the *employee* base class:

```
#include <iostream.h>
#include <string.h>

class employee {
 public:
    employee(char *, char *, float);
    void show_employee(void);
 private:
    char name[64];
    char position[64];
    float salary;
};

employee::employee(char *name, char *position, float salary)
 {
    strcpy(employee::name, name);
    strcpy(employee::position, position);
    employee::salary = salary;
 }

void employee::show_employee(void)
 {
    cout << "Name: " << name << endl;
    cout << "Position: " << position << endl;
    cout << "Salary: $" << salary << endl;
 }
```

```
class manager : public employee {
 public:
  manager(char *, char *, char *, float, float, int);
  void show_manager(void);
 private:
  float annual_bonus;
  char company_car[64];
  int stock_options;
};

manager::manager(char *name, char *position, char *company_car,
    float salary, float bonus, int stock_options) :
    employee(name, position, salary)
 {
    strcpy(manager::company_car, company_car);
    manager::annual_bonus = bonus;
    manager::stock_options = stock_options;
 }

void manager::show_manager(void)
 {
    show_employee();
    cout << "Company car: " << company_car << endl;
    cout << "Annual bonus: $" << annual_bonus << endl;
    cout << "Stock options: " << stock_options << endl;
 }

void main(void)
 {
    employee worker("John Doe", "Programmer", 35000);
    manager boss("Jane Doe", "Vice President", "Lexus",
       50000.0, 5000, 1000);

    worker.show_employee();
    boss.show_manager();
 }
```

CONSTRUCTOR

THIS METHOD WAS INHERITED FROM THE BASE CLASS.

As you can see, the program defines the *employee* base class and then defines the derived *manager* class. Note the *manager* constructor function. When you derive a class from a base class, the derived-class constructor function must invoke the base-class constructor. To invoke the base-class constructor, you place a colon immediately after the derived-class constructor function and then specify the base-class constructor name with the desired parameters, as shown here:

```
manager::manager(char *name, char *position, char *company_car,
    float salary, float bonus, int stock_options) :
    employee(name, position, salary)
 {                                                    Base-class constructor
    strcpy(manager::company_car, company_car);
    manager::annual_bonus = bonus;
    manager::stock_options = stock_options;
 }
```

191

Also, note that the function *show_manager* calls the function *show_employee*, which is a member of the *employee* class. Because the program derives the *manager* class from the *employee* class, the *manager* class can access the *employee* class *public* members, just as if the program defined the members within the *manager* class.

Understanding Inheritance

Inheritance is the ability of a derived class to inherit the characteristics of an existing base class. In simple terms that means that if you have a class whose data or function a new class needs, you can build the new class in terms of the existing (or base) class. The new class, in turn, will inherit the members (characteristics) of the existing class. Using inheritance to build new classes saves you considerable time and programming. Object-oriented programming makes extensive use of inheritance by letting your programs build complex objects out of smaller, more manageable objects.

LOOKING AT A SECOND EXAMPLE

As an example of inheritance, assume that you are using the following *book* base class within an existing program:

```
class book {
 public:
   book(char *, char *, int);
   void show_book(void);
 private:
   char title[64];
   char author[64];
   int pages;
};
```

Next, assume that your program must create a *library_card* class that will add the following data members to the *book* class:

```
char catalog[64];
int checked_out; // 1 if checked out, otherwise 0
```

Your program can use inheritance to derive the *library_card* class from the the *book* class, as shown here:

```
class library_card : public book {
 public:
   library_card(char *, char *, int, char *, int);
   void show_card(void);
 private:
   char catalog[64];
   int checked_out;
};
```

The following program, *BookCard.CPP*, derives the *library_card* class from the *book* class:

```
#include <iostream.h>
#include <string.h>

class book {
 public:
   book(char *, char *, int);
   void show_book(void);
```

```cpp
 private:
  char title[64];
  char author[64];
  int pages;
};

book::book(char *title, char *author, int pages)
 {
   strcpy(book::title, title);
   strcpy(book::author, author);
   book::pages = pages;
 }

void book::show_book(void)
 {
   cout << "Title: " << title << endl;
   cout << "Author: " << author << endl;
   cout << "Pages: " << pages << endl;
 }

class library_card : public book {
 public:
   library_card(char *, char *, int, char *, int);
   void show_card(void);
 private:
   char catalog[64];
   int checked_out;
 };

library_card::library_card(char *title, char *author,
  int pages, char *catalog, int checked_out) :
  book(title, author, pages)
 {
  strcpy(library_card::catalog, catalog);
  library_card::checked_out = checked_out;
 }

void library_card::show_card(void)
 {
   show_book();
   cout << "Catalog: " << catalog << endl;
   if (checked_out)
     cout << "Status: Checked out" << endl;
   else
     cout << "Status: Available" << endl;
 }

void main(void)
 {
   library_card card("Rescued by C++", "Jamsa", 272, "101CPP", 1);

   card.show_card();
 }
```

193

As before, note the *library_card* constructor function calls the *book* class constructor to initialize *book* class members. Also, note the use of the *book* class *show_book* member function within the *show_card* function. Because the *library_card* class inherits the *book* class methods, the *show_card* function can invoke method without having to specify a dot operator, just as if the *library_card* class defined the *show_card* method as one of its own.

UNDERSTANDING PROTECTED MEMBERS

As you examine base-class definitions, you may encounter *public*, *private*, and *protected* class members. As you know, on the one hand a derived class can access base-class *public* members just as if the program defined the base-class members within the derived class. On the other hand, the derived class cannot access base-class *private* members directly. Instead, the derived class must use a base-class interface function to access such members.

A *protected* base-class member falls in between a *private* and *public* member. If a member is *protected*, derived-class objects can access the member as if the member were *public*. To the rest of your program, *protected* members appear as *private*. The only way your program can access *protected* members is to use interface functions. The following *book* class definition uses the *protected* label to let classes derived from the *book* class access the *title*, *author*, and *pages* members directly, using the dot operator:

```
class book {
 public:
  book(char *, char *, int);
  void show_book(void);
 protected:
  char title[64];
  char author[64];
  int pages;
};
```

If you think you may someday derive new classes from a class you are creating, determine whether you want to give those classes direct access to specific members, and then declare those members *protected*, as opposed to *private*.

PROTECTED MEMBERS PROVIDE ACCESS AND SECURITY

As you have learned, *private* class members prevent a program from accessing class members directly. Instead, to access a *private* class member, your program must use an interface function that controls the member's access. When your programs use inheritance, you may find that it simplifies programming if the classes you derive can access base-class members using the dot operator. In such cases, your programs can use *protected* class members. A derived class can access *protected* class members directly using the dot operator. The rest of your program, however, can access the *protected* class members only by using the class interface functions. *Protected* class members sit between *public* class members (which the entire program can access) and *private* class members which only the class itself can access directly.

RESOLVING MEMBER NAMES

When you derive one class from another, there may be times when a class member name in the derived class is the same as a member name in the base class. If such conflicts occur, C++ always uses the derived-class member within the derived-class functions. For example, assume the *book* and *library_card* classes both use the member *price*. In the case of the *book* class, the *price* member corresponds to the book's retail price, such as $29.95. In the case of the *library_card* class, the *price* might include a library discount, such as $24.50. Unless your code tells the *library_card* class functions to do otherwise (via the global resolution operator), the functions will use the derived (*library_card*) class member. Should a *library_card* class function need to access the base (*book*) class price member, the function can use the *book* class name and resolution operator, such as *book::price*. For example, assume the *show_card* function must display both prices. The function would use the following statements:

```
cout << "Library price: $" << price << endl;
cout << "Retail price: $" << book::price << endl;
```

If you execute the program, these statements will display the following output:

```
Library price: $24.50
Retail price: $29.95
```

WHAT YOU MUST KNOW

In this lesson you learned that C++ inheritance capabilities let you build (derive) a new class from an existing class. By building one class from another in this way, you reduce your programming, which, in turn, saves you time. In Lesson 29, "Multiple Inheritance," you will learn that C++ lets you derive a class from two or more base classes. Using multiple base classes to derive a class is called *multiple inheritance*. Before you continue with Lesson 29, however, make sure that you have learned the following key concepts:

- ☑ Inheritance is the ability to derive a new class from an existing base class.

- ☑ The derived class is the new class, and the base class is the original class.

- ☑ When you derive one class from another (the base class), the derived class inherits the base-class members.

- ☑ To derive a class from a base class, you begin your class definition with the *class* keyword, followed by the class name, a colon, and the base class, such as *class dalmatian : dog*.

- ☑ When you derive a class from a base class, the derived class can access the base-class *public* members, just as if the program defined the members within the derived class itself. The derived class must use base-class interface functions to access the base-class *private* data.

- ☑ Within a derived-class constructor function, your program must invoke the base-class constructor by specifying a colon, the function name, and the corresponding parameters immediately after the derived-class constructor function header.

- ☑ To provide derived classes with direct access to specific base-class members, while protecting the members from the remainder of your program, C++ provides *protected* class members. A derived class can access base-class protected members, just as if the members were *public*. To the rest of your program, however, the members appear *private*.

- ☑ If a derived class and base class use the same member name, C++ will use the derived-class member within derived-class functions. If such a function must access the base-class member, you must use the global resolution operator, such as *base_class::member*.

LESSON 29

MULTIPLE INHERITANCE

In Lesson 28, "Understanding Inheritance," you learned that you can build one class from another by letting a class inherit the characteristics of a different class. As it turns out, C++ lets you derive a class from more than one base class. When your class inherits the characteristics of more than one class, you are building the new class using *multiple inheritance*. As you will learn in this lesson, C++ fully supports multiple inheritance. By the time you finish this lesson, you will understand the following key concepts:

- When you derive a class using multiple base classes, you take advantage of *multiple inheritance*.

- Using multiple inheritance, a derived class receives the attributes of two or more base classes.

- When you use multiple inheritance to derive a class, the derived-class constructor must invoke constructor functions for each of the base classes.

- When you derive a class from a derived class, you create an inheritance chain.

Multiple inheritance is a very powerful object-oriented programming tool. Experiment with the programs this lesson presents. As you will find, building one class from an existing class will save you considerable programming.

LOOKING AT A SIMPLE EXAMPLE

As an example of multiple inheritance, assume that you have the following *computer_screen* class:

```
class computer_screen {
 public:
   computer_screen(char *, long, int, int);
   void show_screen(void);
 private:
   char type[32];
   long colors;
   int x_resolution;
   int y_resolution;
};
```

Likewise, assume that you have the following *mother_board* class:

```
class mother_board {
 public:
   mother_board(int, int, int);
   void show_mother_board(void);
 private:
   int processor;
   int speed;
   int RAM;
};
```

Using these two classes, you can derive the *computer* class, as shown here:

```
class computer : public computer_screen, public mother_board
{
 public:
   computer(char *, int, float, char *, long, int, int, int, int, int);
   void show_computer(void);
 private:
   char name[64];
   int hard_disk;
   float floppy;
};
```

As you can see, the class specifies its base classes immediately after the colon that follows the computer class name:

```
class computer : public computer_screen, public mother_board
```
Base classes

The following program, *Computer.CPP*, derives the *computer* class using the *computer_screen* and *mother_board* basc classes:

```
#include <iostream.h>
#include <string.h>

class computer_screen {
 public:
   computer_screen(char *, long, int, int);
   void show_screen(void);
 private:
   char type[32];
   long colors;
   int x_resolution;
   int y_resolution;
};

computer_screen::computer_screen(char *type,
  long colors, int x_res, int y_res)
  {
    strcpy(computer_screen::type, type);
    computer_screen::colors = colors;
    computer_screen::x_resolution = x_res;
    computer_screen::y_resolution = y_res;
  }

void computer_screen::show_screen(void)
  {
    cout << "Screen type: " << type << endl;
    cout << "Colors: " << colors << endl;
    cout << "Resolution: " << x_resolution <<  " by " <<
      y_resolution << endl;
  }
```

197

```cpp
class mother_board {
 public:
   mother_board(int, int, int);
   void show_mother_board(void);
 private:
   int processor;
   int speed;
   int RAM;
};

mother_board::mother_board(int processor, int speed, int RAM)
 {
   mother_board::processor = processor;
   mother_board::speed = speed;
   mother_board::RAM = RAM;
 }

void mother_board::show_mother_board(void)
 {
   cout << "Processor: " << processor << endl;
   cout << "Speed: " << speed << "Mhz" << endl;
   cout << "RAM: " << RAM << " Mb" << endl;
 }

class computer : public computer_screen,
  public mother_board {
 public:
   computer(char *, int, float, char *, long,
     int, int, int, int, int);
   void show_computer(void);
 private:
   char name[64];
   int hard_disk;
   float floppy;
};

computer::computer(char *name, int hard_disk,
  float floppy, char *screen, long colors,
  int x_res, int y_res, int processor, int speed,
  int RAM) : computer_screen(screen, colors, x_res,
  y_res), mother_board(processor, speed, RAM)
 {
   strcpy(computer::name, name);
   computer::hard_disk = hard_disk;
   computer::floppy = floppy;
 }

void computer::show_computer(void)
 {
```

```
   cout << "Type: " << name << endl;
   cout << "Hard disk: " << hard_disk << "Mb" << endl;
   cout << "Floppy disk: " << floppy << "Mb" << endl;
   show_mother_board();
   show_screen();
 }


void main(void)
 {
   computer my_pc("Compaq", 212, 1.44, "SVGA",
     16000000, 640, 480, 486, 66, 8);

   my_pc.show_computer();
 }
```

If you examine the constructor function for the *computer* class, you will find that it invokes the constructor functions for the *mother_board* and *computer_screen* classes, as shown here:

```
computer::computer(char *name, int hard_disk, float floppy,
  char *screen, long colors, int x_res, int y_res,
  int processor, int speed, int RAM) :
  computer_screen(screen, colors, x_res, y_res),
  mother_board(processor, speed, RAM)
```

BUILDING AN INHERITANCE CHAIN

When you use C++ inheritance to derive one class from another, there may be times when you derive a class using a class that the program derives from a base class itself. For example, assume that you need to use the *computer* class as a base class to derive the *work_station* class, shown here:

```
class work_station : public computer {
 public:
   work_station(char *operating_system, char *name,
       int hard_disk, float floppy, char *screen,
       long colors, int x_res, int y_res,
       int processor, int speed, int RAM);
   void show_work_station(void);
 private:
   char operating_system[64];
};
```

The constructor function for the *work_station* class simply invokes the constructor for the *computer* class, which, in turn, invokes the constructors for the *computer_screen* and *mother_board* classes:

```
work_station::work_station(char *operating_system,
  char *name, int hard_disk, float floppy,
  char *screen, long colors, int x_res, int y_res,
  int processor, int speed, int RAM) :
  computer (name, hard_disk, floppy, screen,
```

199

```
      colors, x_res, y_res, processor, speed, RAM)
    {
      strcpy(work_station::operating_system, operating_system);
    }
```

In this example, the *computer* class serves as a base class. However, as you know, the *computer* class was derived from the *computer_screen* and *mother_board* classes. As a result, the *work_station* class inherits the characteristics of all three classes. As Figure 29 shows, the class derivation builds a class chain.

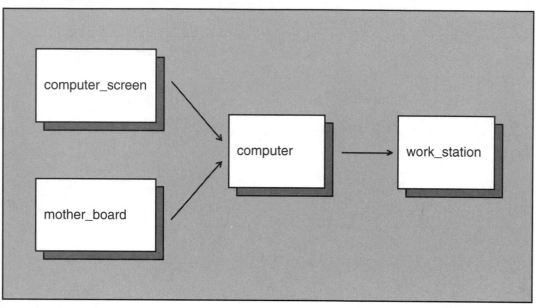

Figure 29 *Building an inheritance chain.*

As your programs begin to make extensive use of inheritance, your inheritance chains and, consequently, the number of members a class inherits can become quite long.

WHAT YOU MUST KNOW

Multiple inheritance is the ability to derive a class from two more or base classes. When you use multiple inheritance, the derived class obtains (inherits) the characteristics (members) of the existing base classes. C++ support for multiple inheritance gives your programs tremendous object-oriented programming capabilities. In Lesson 30, "Private Members and Friends," you will to learn how to provide access to class *private* members to other classes or functions within another class that you have classified as a *friend*. Using such friends, you can grant selected functions direct access to class members while maintaining protection from the rest of your program. Before you continue with Lesson 30, however, make sure you have learned the following key concepts:

☑ Multiple inheritance is the ability of a derived class to inherit the characteristics of two or more base classes.

☑ To derive a class using two or more base classes, you specify the base-class names, separated by commas following the new class name and colon, such as *class cabbit : public cat, public rabbit*.

☑ When you define the derived-class constructor function, you must invoke the constructor functions for each base class, passing to the functions the correct parameters.

☑ When you derive classes, there may be times when the base class you are using is actually derived from other base classes. In such cases, your program creates an inheritance chain. When your program calls the constructor for a derived class, C++ calls the constructor functions for the inherited classes (in succession) as well.

LESSON 30

PRIVATE MEMBERS AND FRIENDS

As you have learned, your programs can only access *private* class members using class-member functions. By using *private* class members as often as possible, you reduce the possibility of a program misusing a member's value because the program can access the member only by using an interface function (which can control the member's access). Depending on your program's object use, however, there may be times when you can significantly increase your program's performance by letting one class directly access the *private* members of another. In this way, your programs avoid the processing overhead (execution time) to invoke the interface functions. For such cases, C++ lets you define a class as a *friend* of another and then lets the *friend* class access the other's *private* class members. This lesson examines how your program specifies that two classes are friends. By the time you finish this lesson, you will understand the following key concepts:

- Using the *friend* keyword, a class can tell C++ who its friends are—in other words, which other classes can access its *private* members directly.

- *Private* class members protect class data; therefore, you should limit your use of *friend* classes to only those classes that have a true need to directly access *private* class members.

- C++ lets you restrict a *friend's* access to a specific set of functions.

Private members let you protect your classes and reduce the possibility of errors. Therefore, you should limit your use of *friend* classes as much as possible. Any time a program can change class members directly, you increase the possibility of errors.

DEFINING A CLASS FRIEND

C++ lets a class access its friend's *private* members. To tell C++ that one class is a *friend* of another, you simply include the *friend* keyword and the corresponding friend's class name within the class definition. For example, the following *book* class states that the *librarian* class is a *friend*. Therefore, objects of the *librarian* class can directly access the *book* class *private* members using the dot operator:

```
class book {
 public:
    book(char *, char *, char *);
    void show_book(void);
    friend librarian;                    // Friend function
 private:
    char title[64];
    char author[64];
    char catalog[64];
};
```

As you can see, to specify a *friend* only requires one statement within the class definition. For example, the following program, *ViewBook.CPP*, uses the *librarian* as a *friend* of the *book* class. Therefore, the functions within the *librarian* class can directly access *book* class *private* members. In this example, the program uses the *librarian* class *change_catalog* function to change the card catalog number for a specific book:

201

```
#include <iostream.h>
#include <string.h>

class librarian;

class book {
 public:
   book(char *, char *, char *);
   void show_book(void);
   friend librarian;                          // Friend function
 private:
   char title[64];
   char author[64];
   char catalog[64];
};

book::book(char *title, char *author, char *catalog)
 {
   strcpy(book::title, title);
   strcpy(book::author, author);
   strcpy(book::catalog, catalog);
 }

void book::show_book(void)
 {
   cout << "Title: " << title << endl;
   cout << "Author: " << author << endl;
   cout << "Catalog: " << catalog << endl;
 }

class librarian {
 public:
   void change_catalog(book *, char *);
   char *get_catalog(book);
};

void librarian::change_catalog(book *this_book,
  char *new_catalog)
 {
   strcpy(this_book->catalog, new_catalog);
 }

char *librarian::get_catalog(book this_book)
 {
   static char catalog[64];

   strcpy(catalog, this_book.catalog);
   return(catalog);
 }

void main(void)
 {
```

```
    book programming("Rescued By C++, Third Edition", "Jamsa", "P101");
    librarian library;

    programming.show_book();
    library.change_catalog(&programming, "EASY C++ 101");
    programming.show_book();
}
```

As you can see, the *ViewBook.CPP* program passes the *book* object to the *librarian* class *change_catalog* function by address. Because the function is changing a member of the class, the program must pass the parameter by address and then use a pointer to access the member. Experiment with this program by removing the *friend* entry from the *book* class definition. Because the *librarian* class no longer has access to the *book* class *private* members, the C++ compiler will generate syntax errors each time the program tries to reference the *book* class *private* data.

Understanding Class Friends

Normally, the only way your programs can access class *private* data is to use the class interface functions. Depending on your program's object use, there may be times when it is convenient (or more efficient in terms of processing speed) to grant one class access to the *private* members of another. To do so, you must inform the C++ compiler that the class is a *friend*. The compiler, in turn, will let the *friend* class access the *private* members as required. To define a class as a *friend*, place the C++ *friend* keyword and the corresponding class name within the *public* members of a class definition, as shown here:

```
class abbott {
  public:
     friend costello;
     // Other members
  private:
     // Private members
};
```

How Friends Differ from Protected Members

In Lesson 28 "Understanding Inheritance," you learned that C++ supports *protected* members, which let derived classes access the *private* members of a base class directly, using the dot operator. Remember that the only classes that can access *protected* class members are classes that your program derives from the base class—in other words, classes that inherited the base-class members. C++ *friend* classes are normally unrelated classes (that is, one class does not inherit members from another). The only way such an unrelated class can access the *private* class members of a second class is if the second class tells the compiler that the unrelated class is a *friend*.

RESTRICTING A FRIEND'S ACCESS

As you just learned, when you specify one class as a *friend* of another, you provide the class with access to its friend's private data. However, the more access you give to class *private* data, the greater your chance of introducing program errors. Therefore, if only a few functions within a *friend* class must access the *private* data of another class, C++ lets you specify that the program only lets specific *friend* class functions access the *private* members. Assume, for example, that the *librarian* class, presented in the previous program, has many different class functions. Likewise, assume that only the functions *change_catalog* and *get_catalog* require access to the *book* class *private* data. Within the *book* class definition, you can restrict access to *private* members to only those two functions, as shown here:

```
class book {
 public:
   book(char *, char *, char *);
   void show_book(void);
   friend char *librarian::get_catalog(book);
   friend void librarian::change_catalog(book *, char *);
 private:
   char title[64];
   char author[64];
   char catalog[64];
};
```

As you can see, the *friend* statements include complete function prototypes for each of the *friend* class functions that can access the *private* members directly.

Understanding Friend Functions

When your program uses friends to access class *private* data, you can restrict the number of class-member functions within the *friend* class that can access the *private* data by using *friend functions*. To declare a *friend* function, you specify the *friend* keyword, followed by a complete prototype for the function that must access the data, as shown here:

```
public:
   friend class_name::function_name(parameter types);
```

Only the member functions you specify as *friends* can access the class *private* members directly, using the dot operator.

When your programs begin to reference one class from another, you can encounter syntax errors if the order of your class definitions is not correct. In this case, the *book* class definition uses function prototypes defined for the *librarian* class. Therefore, the *librarian* class definition must precede that of the *book* class. However, if you examine the *librarian* class, you will find that it references the *book* class, as shown here:

```
class librarian {
 public:
   void change_catalog(book *, char *);
   char *get_catalog(book);
};
```

Because you cannot precede this *library* class definition with the *book* class while preceding the *book* class definition with the *library* class definition, C++ lets you use a one-line class definition that tells the compiler that *book* is a class that your program will define later:

```
class book;
```

204 The following program, *LimitFri.CPP*, uses *friend* functions to limit *librarian* class access to *book* class *private* data. Note the order of the class definitions:

```
#include <iostream.h>
#include <string.h>

class book;

class librarian {
 public:
   void change_catalog(book *, char *);
   char *get_catalog(book);
};

class book {
 public:
   book(char *, char *, char *);
   void show_book(void);
   friend char *librarian::get_catalog(book);
   friend void librarian::change catalog(book *, char *);
 private:
   char title[64];
   char author[64];
   char catalog[64];
};

book::book(char *title, char *author, char *catalog)
 {
   strcpy(book::title, title);
   strcpy(book::author, author);
   strcpy(book::catalog, catalog);
 }

void book::show_book(void)
 {
   cout << "Title: " << title << endl;
   cout << "Author: " << author << endl;
   cout << "Catalog: " << catalog << endl;
 }

void librarian::change_catalog(book *this_book,
  char *new_catalog)
 {
   strcpy(this_book->catalog, new_catalog);
 }

char *librarian::get_catalog(book this_book)
 {
   static char catalog[64];

   strcpy(catalog, this_book.catalog);
   return(catalog);
 }

void main(void)
 {
```

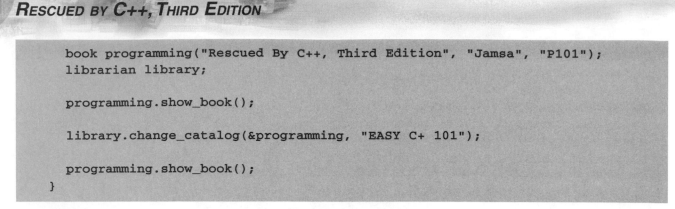

```
      book programming("Rescued By C++, Third Edition", "Jamsa", "P101");
      librarian library;

      programming.show_book();

      library.change_catalog(&programming, "EASY C+ 101");

      programming.show_book();
   }
```

As you can see, the *LimitFri.CPP* program first uses a single-line entry that tells the compiler that it will later define the *book* class. Because the entry makes the compiler aware of the *book* class, the *librarian* class definition can make references to the *book* class, which the program does not define until later.

Introducing a Class Identifier

An identifier is a name, such as a variable or a class type. When your programs use *friend* classes, there may be times when one class definition refers to a second class (its name or identifier) that the C++ compiler does not yet know about. In such cases, the C++ compiler will generate syntax-error messages. To eliminate such "who gets defined first" errors, C++ lets you include a single-line entry, shown here, near the top of your source code that introduces a class identifier:

```
class class_name;
```

The entry tells the C++ compiler that your program will later define the class specified, but for now, it's OK for your program to reference the class name.

WHAT YOU MUST KNOW

In this lesson, you learned how to use *friend* classes to let one class directly access the *private* members of another class, using the dot operator. In Lesson 31, "Using Function Templates," you will learn how to use C++ function templates to simplify definitions of similar functions. Before you continue with Lesson 31, however, make sure that you have learned the following key concepts:

- ☑ Using C++ friends, your programs can let one class directly access the *private* members of another class, using the dot operator.
- ☑ To declare one class as a *friend* of a second class, you specify the *friend* keyword, followed by the class name, within the second class definition.
- ☑ After you declare a class as a *friend* to a second class, all member functions within the *friend* class can access the *private* members within the second class.
- ☑ To limit the number of *friend* methods that can access class *private* data, C++ lets you specify *friend* functions. To declare a *friend* function, you specify the *friend* keyword, followed by the function prototype for the function that needs to access class *private* members.
- ☑ When you declare friend functions, you may encounter syntax errors if the order in which your program defines classes is not correct. If you must tell C++ that an identifier is a class that your program defines later, you can include a single-line entry similar to the *class class_name;* statement.

LESSON 31

USING FUNCTION TEMPLATES

When you create functions in your programs, there will be times when two functions perform similar processing, but work with different data types (for example, one function uses parameters of type *int* and the other type *float*). As you learned in Lesson 15, "Overloading Functions," by overloading functions you can use the same function name to handle operations that use different parameter types. When the functions return different types, however, you must define functions with unique names. For example, assume you have a function named *max* that returns the maximum of two integer values. If you later require a similar function that returns the maximum of two floating-point values, you must define a different function, such as *fmax*. In this lesson you will learn how to use C++ *templates* to quickly create functions that return different value types. By the time you finish this lesson, you will understand the following key concepts:

- *Templates* define a set of statements with which your programs can later create more functions.

- Programs often use *function templates* to quickly define two or more functions that work with similar statements but with different parameter or return value types.

- Function templates have specific names that match the function name you want to use within your program.

- After your program defines a function template, your program can later create a function by using the template to specify a prototype that includes the template name, the function's return value, and the function's parameter types.

- During compilation, the C++ compiler will create the functions within your program using the types your program specifies in prototypes that use the template name.

Function templates use a unique syntax that you may find intimidating. However, after you use one or two templates, you will find they are actually very easy to use.

Note: *The* **Turbo C++ Lite** *compiler provided on this book's companion CD-ROM does not support C++ templates.*

CREATING A SIMPLE FUNCTION TEMPLATE

A *function template* defines a typeless (or generic) function with which your programs later create a function that uses the types the program requires. For example, the following statements define a template for a function named *max*, which returns the larger of two values:

```
template<class T> T max(T a, T b)
  {
    if (a > b)
      return(a);
    else
      return(b);
  }
```

In this example, the letter *T* is the template's generic type. After you define the template within your program, you declare function prototypes for each type you require. In the case of the *max* template, the following prototypes create functions of type *float* and *int*:

```
float max(float, float);
int max(int, int);
```

When the C++ compiler encounters these function prototypes within your program, the compiler will replace the template's type *T* with the corresponding type as it builds the functions. In the case of the *float* implementation of *max*, the compiler's substitution becomes the following:

```
float max(float, float);

            template<class T> T max(T a, T b)
              {
                if (a > b)
                    return(a);
                else
                    return(b);
              }

                        float max(float a, float b)
                          {
                            if (a > b)
                                return(a);
                            else
                                return(b);
                          }
```

The following program, *Max_Temp.CPP*, uses the *max* template to create functions of type *int* and *float*:

```
#include <iostream.h>

template<class T> T max(T a, T b)
  {
    if (a > b)
      return(a);
    else
      return(b);
  }

float max(float, float);
int max(int, int);

void main(void)
  {
    cout << "The maximum of 100 and 200 is " <<
      max(100, 200) << endl;

    cout << "The maximum of 5.4321 and 1.2345 is " <<
      max(5.4321, 1.2345) << endl;
  }
```

During compilation, the C++ compiler automatically creates the statements that build one function that supports the type *int*, and a second function that supports the type *float*. Because the C++ compiler manages the statements that correspond to functions you create using templates, the compiler lets you use the same function name for functions that return different types, something you could not do using only function overloading (as discussed in Lesson 15).

Using Function Templates

As your programs become more complex, there will be many times when you need similar functions that perform the same operation, but on different data types. Function templates let your programs define a generic or typeless function. When your program must use the function for a specific type, such as *int* or *float*, your program specifies a function prototype that uses the template function name and that specifies the function's return value and parameter types. During compilation, the C++ compiler will then create the corresponding function. By creating function templates, you reduce the number of functions you must code yourself, and your programs can use the same function name for functions that perform a specific operation, regardless of the function's return value and parameter types.

TEMPLATES THAT USE MULTIPLE TYPES

The previous template definition for the *max* function used only the generic type *T*. In many cases, however, a template function may need multiple types. For example, the following statements create a template for the function *show_array*, which displays an array's elements. The template uses the type *T* to define the array type and the type *T1* to specify the parameter count type:

```
template<class T, class T1> void show_array(T *array, T1 count)
  {
    T1 index;
    for (index = 0; index < count; index++)
      cout << array[index] << ' ';

    cout << endl;
  }
```

As before, the program must specify function prototypes that correspond to each function types:

```
void show_array(int *, int);
void show_array(float *, unsigned);
```

The following program, *Show_Tem.CPP*, uses the *show_array* template to create functions that display arrays of type *int* and type *float*:

```
#include <iostream.h>

template<class T, class T1> void show_array(T *array, T1 count)
  {
    T1 index;

    for (index = 0; index < count; index++)
      cout << array[index] << ' ';

    cout << endl;
  }

void show_array(int *, int);
void show_array(float *, unsigned);
```

209

```
void main(void)
 {
   int pages[] = { 100, 200, 300, 400, 500 };
   float prices[] = { 10.05, 20.10, 30.15 };

   show_array(pages, 5);
   show_array(prices, 3);
 }
```

Templates and Multiple Types

As your function templates become more complex, they may provide support for multiple types. For example, your program might create a template for a function named *array_sort*, which sorts the elements of an array. In this case, the function may use two parameters, the first of which corresponds to the array, and the second to the number of elements in the array. If the program assumes the array will never hold more than 32,767 values, the program can use a parameter of type *int* for the array size parameter. However, a more generic template would let the program specify its own type for the parameter, as shown here:

```
template<class T, class T1> void array_sort(T array[], T1 elements)
 {
    // statements
 }
```

Using the *array_sort* template, your program can create functions that sort a small array of type *float* (less than 128 elements) and a very large array of type *int* using the following prototypes:

```
void array_sort(float, char);
void array_sort(int, long);
```

WHAT YOU MUST KNOW

Using function templates reduces your programming by letting the C++ compiler generate the statements for functions that differ only in return or parameter types. In Lesson 32, "Using Class Templates," you will learn how to use templates to create typeless or generic classes. Before you continue with Lesson 32, however, make sure that you have learned the following key concepts:

- ☑ Function templates let you declare a typeless or generic function.

- ☑ When your program must use the function with specific data types, your program specifies a function prototype that defines the correct types.

- ☑ When the C++ compiler encounters the function prototype, the compiler will create the corresponding function statements, substituting in the correct types.

- ☑ Your programs should create templates for common functions that work with various types. In other words, if you use only one type with a function, you do not need to use template.

- ☑ If a function requires more than one type, the template simply assigns each type a unique identifier, such as *T*, *T1*, or *T2*. Later, during compilation, the C++ compiler will assign the correct types you specify using the function prototype.

LESSON 32

USING CLASS TEMPLATES

In Lesson 31, "Using Function Templates," you learned how to use C++ function templates to create generic or typeless functions. By defining the function templates, you can later direct the C++ compiler to create functions for you, which differ by return value or parameter type. Just as there may be times when you create similar functions that only differ by type, there may be times when your programs must create generic classes. To do so, your programs can define *class templates*. This lesson examines the steps your programs must perform to define and use class templates. By the time you finish this lesson, you will understand the following key concepts:

- Using the *template* keyword and type symbols such as *T, T1,* or *T2,* your programs can create a class template—the template class definition can use these symbols to specify member data, member function return and parameter value types, and so on.

- To create class objects using class templates, your programs simply reference the class name followed by the types the compiler is to assign to each type symbol. The program specifies types within angle brackets (such as *<int, float>*) and the variable name.

- If the class provides a constructor function with which you can initialize member variables, you can invoke the constructor when you create an object using the template, such as *class_name<int, float> values(200);*.

- When the C++ compiler encounters the object declaration, the compiler will create a class from the template that uses the corresponding types.

As was the case with function templates, you may find class templates intimidating "at first glance." However, after you create and use one or two class templates, you will find they are very straightforward and easy to use.

Note: The Turbo C++ Lite compiler provided on this book's companion CD-ROM does not support C++ templates.

CREATING A CLASS TEMPLATE

As an example of how to use a class template, assume you are creating an *array* class that provides methods that calculate the sum and average of values stored in an array. Provided you are working with an array of type *int,* your class might appear as follows:

```
class array {
 public:
   array(int size);
   long sum(void);
   int average_value(void);
   void show_array(void);
   int add_value(int);
 private:
   int *data;
   int size;
   int index;
 };
```

211

The following program, *I_Array.CPP*, uses the *array* class to work with values of type *int*:

```
#include <iostream.h>
#include <stdlib.h>

class array {
 public:
   array(int size);
   long sum(void);
   int average_value(void);
   void show_array(void);
   int add_value(int);
 private:
   int *data;
   int size;
   int index;
};

array::array(int size)
 {
   data = new int[size];

   if (data == NULL)
    {
      cerr << "Insufficient memory--program ending" << endl;
      exit(1);
    }

   array::size = size;
   array::index = 0;
 }

long array::sum(void)
 {
   long sum = 0;

   for (int i = 0; i < index; i++)
     sum += data[i];

   return(sum);
 }

int array::average_value(void)
 {
   long sum = 0;

   for (int = 0; i < index; i++)
     sum += data[i];

   return(sum / index);
 }

void array::show_array(void)
 {
```

```
      for (int i = 0; i < index; i++)
        cout << data[i] << ' ';

      cout << endl;
  }


  int array::add_value(int value)
   {
     if (index == size)
       return(-1);          // Array is full
     else
      {
        data[index] = value;
        index++;
        return(0);          // Successful
      }
   }


  void main(void)
   {
     array numbers(100);   // 100 element array
     int i;

     for (i = 0; i < 50; i++)
       numbers.add_value(i);

     numbers.show_array();

     cout << "The sum of the numbers is " << numbers.sum() << endl;
     cout << "The average value is " << numbers.average_value() << endl;
   }
```

As you can see, the *I_Array.CPP* program allocates a 100-element array and then assigns 50 elements to the array using the *add_value* method. Within the array class, the *index* member tracks the number of elements the array currently stores. If the program tries to add more elements than the array can hold, the *add_value* function will return an error. As you can see, the *average_value* function uses the *index* member to determine the array's average value. The program allocates memory for the array using the *new* operator, which Lesson 33, "Using the C++ Free Store," discusses in detail.

Understanding Class Templates

As the number of classes you create increases, you will find that a class that you have created for one program, is very similar to a class you need for your current program. In many cases, the classes may differ only by type. In other words, one class may support integer values while the class your program requires must support values of type *float*. To increase your ability to reuse existing code, C++ lets your programs define class templates. In short, a class template defines a typeless (or generic) class with which you can later create objects that require a specific type. When the C++ compiler encounters an object declaration that your program bases on a class template, the compiler will use the types the declaration specifies to build the necessary class type. By letting you quickly create classes that differ only by type, class templates reduce your programming, which, in turn, will save you time.

As a second example of using a class template, assume that your program now must work with a floating-point array in addition to the integer array. One way to provide support for the different array types is to create different classes. Using a class template, on the other hand, you can eliminate the need to duplicate the class. The following class template creates the generic *array* class:

```
template<class T, class T1>
class array {
 public:
    array(int size);          ◄——— CONSTRUCTOR – SAME
    T1 sum(void);                                  NAME
  ► T average_value(void);                         AS
    void show_array(void);                         CLASS
    int add_value(T);                              NAME
 private:
    T *data;
    int size;
    int index;
};
```

The *template* statement defines the type symbols *T* and *T1*. In the case of an array of integer values, *T* will correspond to *int*, and *T1* to *long*. Likewise, for an array of floating-point values *T* and *T1* are both *float*. Take time now to ensure you understand how the C++ compiler will later substitute the types you specify for the symbols *T* and *T1*.

Before each class function, you must specify the same *template* entry. In addition, immediately following the class name you must include the type classes, such as *array<T, T1>::average_value*. The following statements, for example, illustrate the class *average_value* function definition:

```
template<class T, class T1>
T array<T, T1>::average_value(void)
  {
    T1 sum = 0;
    int i;
    for (i = 0; i < index; i++)
      sum += data[i];
    return(sum / index);
  }
```

After the template exists, you can create a class of the correct type by specifying the class name, followed by the desired type within angle brackets, as shown here:

```
array<int, long>  numbers(100);        Template name
array<float, float>  values(200);      Template types
```

The program *GenArray.CPP*, uses the array class template to create two classes: one that supports values of type *int*, and one that supports values of type *float*:

```
#include <iostream.h>
#include <stdlib.h>

template<class T, class T1>
class array {
```

```
  public:
    array(int size);
    T1 sum(void);
    T average_value(void);
    void show_array(void);
    int add_value(T);
  private:
    T *data;
    int size;
    int index;
  };

template<class T, class T1>
array<T, T1>::array(int size)
  {
    data = new T[size];

    if (data == NULL)
     {
       cerr << "Insufficient memory--program ending" << endl;
       exit(1);
      }

    array::size = size;
    array::index = 0;
  }

template<class T, class T1>
T1 array<T, T1>::sum(void)
  {
    T1 sum = 0;

    for (int i = 0; i < index; i++)
      sum += data[i];

    return(sum);
  }

template<class T, class T1>
T array<T, T1>::average_value(void)
  {
    T1 sum = 0;

    for (int i = 0; i < index; i++)
      sum += data[i];

    return(sum / index);
  }
```

```
template<class T, class T1>
void array<T, T1>::show_array(void)
 {

   for (int i = 0; i < index; i++)
     cout << data[i] << ' ';

   cout << endl;
 }


template<class T, class T1>
int array<T, T1>::add_value(T value)
 {
   if (index == size)
     return(-1);          // Array is full
   else
    {
      data[index] = value;
      index++;
      return(0);        // Successful
    }
 }

void main(void)
 {
   // 100 element array
   array<int, long> numbers(100);
   // 200 element array
   array<float, float> values(200);
   int i;

   for (i = 0; i < 50; i++)
     numbers.add_value(i);

   numbers.show_array();

   cout << "The sum of the numbers is " << numbers.sum() << endl;
   cout << "The average value is " << numbers.average_value() << endl;

   for (i = 0; i < 100; i++)
     values.add_value(i * 100);

   values.show_array();

   cout << "The sum of the numbers is " << values.sum() << endl;
   cout << "The average value is " << values.average_value() << endl;
 }
```

216   The best way to understand class templates is to print two copies of this program. In the first printout, replace each occurrence of *T* and *T1* with *int* and *long*. In the second print, replace *T* and *T1* with *float*.

Declaring Objects That Are Based on a Class Template

To create objects using a class template, you simply specify the template class name, followed by left-and-right angle brackets that specify types you want the compiler to substitute for the symbols *T, T1, T2,* and so on. Next, your program specifies the object (variable) name with the parameter values you want to pass to the class constructor function, as shown here:

```
template_class_name<type1, type2> object_name(parameter1, parameter2);
```

When the C++ compiler encounters the declaration, it will create a class based on the types your program specified. The following statement, for example, uses the *array* class template to create an array of type *char* that can store 100 elements:

```
array<char, int> small_numbers(100);
```

WHAT YOU MUST KNOW

In this lesson you learned that class templates help you reduce duplicate program code when you need similar class objects that differ only by type. Because class templates can be complex, they can become intimidating. When you define your class, begin your definition as if you were creating the class for a specific type. After you have defined the entire class, determine which members must change to support different object types. Next, replace those members' types with symbols such as *T, T1, T2,* and so on.

The programs this lesson presented used the C++ *new* operator to allocate memory dynamically (as the program executed) to store the arrays. In Lesson 33, "Using the C++ Free Store," you will examine the *new* operator in detail. Before you continue with Lesson 33, however, make sure that you have learned the following key concepts:

- ☑ Class templates let you eliminate duplicate code for classes whose objects differ only by member type.

- ☑ To create a class template, precede the class definition with the *template* keyword and the type symbols, such as *T* or *T1*.

- ☑ You must precede each class function definition with the same *template* statement. In addition, you must specify the template types within right-and-left angle brackets between the class name and resolution operator, such as *class_name<T, T1>::function_name*.

- ☑ To create a class using the *template* keyword, you specify the class name, followed by left-and-right angle brackets that contain the replacement types, such as *class_name<int, long> object*.

217

ICQ VISUAL C++/MFC USERS

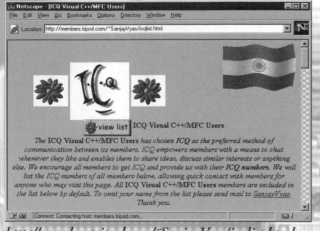

http://member.tripod.com/~SanjayVyas/icqlist.html

AVAILABLE C++ LIBRARIES

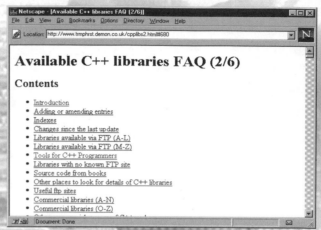

http://www.trmphrst.demon.co.uk/cpplibs2.html#680

NERD WORLD: C++ PROGRAMMING

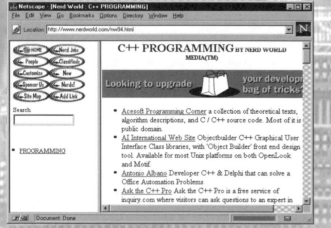

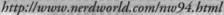

http://www.nerdworld.com/nw94.html

C&C++ RESOURCES

http://www.eecs.utoledo.edu/~cwinner/c.html

TOOLS FOR C++ PROGRAMMERS

http://www.trmphrst.demon.co.uk/cpplibs4.html—
#Tools_for_C++_Programmers

DOCUMENTATION SYSTEM FOR C/C++

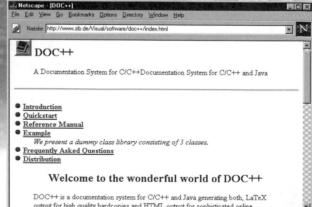

http://www.zib.de/Visual/software/doc++/index.html

ADVANCED C++ CONCEPTS

This book's first five sections have presented the C++ and object-oriented programming concepts you must know to create powerful programs. In this section, you will round out your C++ knowledge as you learn to allocate memory during your program's execution. You will also learn how to perform file operations within C++ and how to use C++ exceptions to respond to unexpected errors within your programs. As you read through this section's lessons, don't let the section's title, "Advanced C++ Concepts," intimidate you. As you will find, the concepts are no more difficult than those you have mastered throughout this book. The lessons in this section include:

SECTION SIX

LESSON 33

USING THE C++ FREE STORE

As you have learned, when your program declares an array, the C++ compiler allocates memory to store the array elements. Over time, it is possible that the size of your array will not be large enough to hold the necessary data. For example, assume that you create an array to hold 100 stock prices. If you later must store more than 100 stock prices, you must edit and recompile your program. As an alternative to allocating fixed-size arrays, your programs can *dynamically* allocate the amount of memory they require as they execute. For example, if your program must track 100 stocks, it would allocate enough memory to store 100 array elements. Likewise, if your program must track only 25 stocks, it could allocate less memory. By dynamically allocating memory, your programs continue to change to meet your needs, without the need for additional programming. When your programs allocate memory during execution, they specify the amount of memory they require and C++ returns a pointer to the memory. C++ allocates memory from a region of memory called the *free store*. This lesson examines the steps your programs must perform to dynamically allocate and later release memory as they execute. By the time you finish this lesson, you will understand the following key concepts:

- To allocate memory during execution, C++ programs use the *new* operator.

- Using the *new* operator, programs specify the amount of memory they require in bytes.

- If the *new* operator successfully allocates the specified memory, *new* will return a pointer to the start of the memory region.

- If the *new* operator cannot satisfy your program's memory request (there may be no free memory left), *new* will return a NULL pointer.

- To later release the memory the program has allocated using *new*, programs use the C++ *delete operator*.

Dynamically allocating memory during execution is a very powerful capability. Experiment with the programs this lesson presents. As you will learn, dynamic memory allocation is actually very easy.

USING THE NEW OPERATOR

The C++ *new* operator lets your programs dynamically allocate memory as they execute. To use the *new* operator you must specify the number of bytes your program requires. For example, assume that your program requires a 50-byte array. Using the *new* operator, you can allocate the memory as shown here:

```
char *buffer = new char[50];
```

If the *new* operator successfully allocates memory, *new* will return a pointer to the start of the memory region. In this case, because the program is allocating memory to store a character array, the program assigns the address to a pointer variable to the type *char*. If the *new* operator cannot allocate the memory the program requests, *new* will return a NULL pointer. Each time your program allocates memory dynamically using the *new* operator, your program should test *new's* return value to determine if it is NULL. If the *new's* return value is NULL, *new* could not fulfill the program's memory request.

Why Programs Dynamically Allocate Memory Using new

Many programs make extensive use of arrays to store multiple values of a specific type. As programmers design their programs, they will normally try to declare arrays whose size is large enough to meet the program's future needs. Unfortunately, should the program's memory needs later exceed those the programmer anticipated, someone must edit and recompile the program.

As an alternative to editing and recompiling programs that simply require more memory, you should create your programs so they dynamically allocate the memory they require as they execute, using the *new* operator. In this way, your programs can modify their memory use as your needs change—eliminating your need to edit and recompile the program.

As an example of dynamic-memory allocation, the following program, *Use_New.CPP*, uses the *new* operator to allocate a pointer to a 100-character array:

```
#include <iostream.h>

void main(void)
 {
   char *pointer = new char[100];

   if (pointer != NULL)
     cout << "Memory successfully allocated" << endl;
   else
     cout << "Error allocating memory" << endl;
 }
```

As you see, the *Use_New.CPP* program immediately tests the value the *new* operator assigned to the variable pointer. If pointer contains the value NULL, *new* was unable to allocate the requested memory. If the pointer does not contain NULL, *new* successfully allocated the memory and the pointer contains the address of the start of the memory block.

If new Cannot Allocate the Requested Memory, new Will Return NULL

When your program uses the *new* operator to allocate memory, there may be times when C++ cannot fulfill the program's memory request because there is not enough available memory in the free store. If the *new* operator is unable to allocate the requested memory, *new* assigns the pointer the value NULL. By testing the pointer's value, your program can determine if *new* satisfied the memory request. The following statement, for example, uses *new* to allocate memory for an array of 500 floating-point values:

```
float *array = new float[500];
```

To determine if *new* successfully allocated the memory, your program should compare the pointer value to NULL, as shown here:

```
if (array != NULL)
   cout << "Successful memory allocation" << endl;
else
   cout << "new did not allocate memory" << endl;
```

The previous program used the *new* operator to allocate 100 bytes of memory. Because the program "hard codes" (specifies within the code) the amount of memory it requires, you must edit the program and recompile if you want the program to allocate more or less memory. As briefly discussed, one reason to dynamically allocate memory is to eliminate your need to edit and recompile as the program's memory needs change.

The following program, *Ask_Mem.CPP*, prompts the user for the number of bytes of memory the program will require and then allocates the memory using the *new* operator:

```cpp
#include <iostream.h>

void main(void)
  {
    int size;
    char *pointer;

    cout << "Type in the array size, up to 30000: ";
    cin >> size;

    if (size <= 30000)
      {
        pointer = new char[size];

        if (pointer != NULL)
          cout << "Memory successfully allocated" << endl;
        else
          cout << "Unable to allocate memory" << endl;
      }
  }
```

When your programs use the *new* operator to dynamically allocate memory, your programs will usually have a way to determine, on their own, how much memory to allocate. For example, if your program allocates memory to hold information about employees, your program might store the number of employees within a file. When the program starts, it can read the number of employees from the file and then allocate the corresponding memory.

The following program, *NoMemory.CPP*, allocates memory 10,000 characters at a time until *new* cannot allocate any more memory from the free store. In other words, the program keeps allocating memory until it uses up the free store's available memory. When the program successfully allocates memory, the program displays a message stating it allocated the memory. When the memory allocation fails, the program displays an error message and ends, as shown here:

```cpp
#include <iostream.h>

void main(void)
  {
    char *pointer;

    do {
      pointer = new char[10000];

      if (pointer != NULL)
        cout << "Allocated 10,000 bytes" << endl;
```

```
        else
          cout << "Memory allocation failed" << endl;

      } while (pointer != NULL);
  }
```

Note: *If you compile and execute in the MS-DOS environment, you may be surprised that the free store runs out of memory after your program's memory allocation exceeds 64Kb. By default, most MS-DOS-based C++ compilers use a small memory model, which provides a 64Kb free store. In a similar way, if you are using the MS-DOS environment, the largest memory region your programs can access might be restricted to 64Kb. To learn how to increase the available free store size, refer to the book* **Jamsa's C/C++ Programmer's Bible***, Jamsa Press, 1997.*

Understanding the Free Store

Each time your program executes, the C++ compiler sets aside a region of unused memory called the *free store*. Using the *new* operator, your program can allocate memory from the free store during execution. By using the free store to allocate memory in this way, your programs are not constrained by fixed array sizes. Depending on your operating system and compiler memory model, the size of the free store will differ. As the amount of dynamic memory your programs require increases, you must be aware of your system's free-store limitations.

RELEASING MEMORY WHEN YOUR PROGRAM NO LONGER NEEDS IT

As you have learned, the C++ *new* operator lets your programs allocate memory dynamically during execution. When your program no longer needs the allocated memory, your program should release the memory using the *delete* operator. To release memory using the *delete* operator, you simply specify a pointer to the memory region, as shown here:

```
delete pointer;
```

The following program, *Del_Mem.CPP*, uses the *delete* operator to release memory the *new* operator allocated:

```
#include <iostream.h>
#include <string.h>

void main(void)
  {
    char *pointer = new char[100];

    strcpy(pointer, "Rescued by C++");

    cout << pointer << endl;

    delete pointer;
  }
```

By default, if your programs do not delete their allocated memory, the operating system automatically releases the memory after the program ends. However, if your program uses the *delete* operator to release memory as soon as it no longer needs the memory, the program makes the memory available for another use (for example, your program may use the *new* operator to allocate the memory for a different use or the operating system may allocate the memory for use by a different program).

LOOKING AT A SECOND EXAMPLE

The following program, *AloccArr.CPP*, allocates memory to store an array of 1,000 integer values. The program then assigns the array the values 1 though 1,000, displaying the values on the screen. Next, the program releases the memory and allocates a 2,000-byte floating-point array, assigning the values 1.0 through 2,000.0 to the array, as shown here:

```cpp
#include <iostream.h>

void main(void)
  {
    int *int_array = new int[1000];
    float *float_array;
    int i;

    if (int_array != NULL)
      {
        for (i = 0; i < 1000; i++)
          int_array[i] = i + 1;

        for (i = 0; i < 1000; i++)
          cout << int_array[i] << ' ';

        delete int_array;
      }

    float_array = new float[2000];

    if (float_array != NULL)
      {
        for (i = 0; i < 2000; i++)
          float_array[i] = (i + 1) * 1.0;

        for (i = 0; i < 2000; i++)
          cout << float_array[i] << ' ';

        delete float_array;
      }
  }
```

As a rule, your programs should release memory using the *delete* operator as soon as they are done using the memory.

WHAT YOU MUST KNOW

In this lesson you learned that your programs can dynamically allocate memory as they execute from a collection of unused memory called the free store, using the *new* operator. By dynamically allocating memory in this way, your programs change as your needs change, without requiring you to edit and recompile the program. In Lesson 34, "Controlling Free Store Operations," you will learn how to control the free-store memory allocation and the operations *new* performs when it cannot satisfy a memory request. Before you continue with Lesson 34, however, make sure that you have learned the following key concepts:

☑ A program's ability to dynamically allocate memory as it executes eliminates the program's dependence on fixed-size arrays.

☑ If the *new* operator successfully allocates the memory your program requests, *new* will return a pointer to the start of the memory region.

☑ If *new* cannot successfully allocate the memory your program requests, *new* will return a NULL pointer.

☑ Each time your programs dynamically allocate memory using *new*, your programs should test the pointer value *new* returns to determine if it is NULL, which indicates *new* was unable to allocate the memory.

☑ Using a pointer or an array, your programs can access the memory they allocate using the *new* operator.

☑ The *new* operator allocates memory from a block of unused memory called the free store.

☑ Depending on your operating system and compiler memory model, the size of the free store will differ. The MS-DOS environment, for example, may limit the free store to 64Kb.

☑ When your program no longer requires its allocated memory, your program should release the memory back to the free store using the *delete* operator.

LESSON 34

CONTROLLING FREE STORE OPERATIONS

In Lesson 33, "Using the C++ Free Store," you learned that as your programs execute, they can use the *new* operator to dynamically allocate memory from the free store. If the *new* operator successfully allocates the memory, your program will receive a pointer to the memory range. If *new* cannot allocate the memory the program requests, *new* will return the NULL value. Depending on your program's purpose, there may be times when you want the program to perform specific operations when *new* cannot satisfy a memory request. In this lesson, you will learn how to direct C++ to call a specific function within your program when *new* cannot satisfy a memory request. By the time you finish this lesson, you will understand the following key concepts:

- Within your program, you can define your own *insufficient-memory handler* (a function) that C++ invokes when *new* cannot satisfy a memory request.

- C++ lets you define your own *new* operator to allocate and possibly initialize memory.

- C++ lets you define your own *delete* operator to discard memory.

As you will learn, by creating custom *new* and *delete* operators, you can better control how your programs handle insufficient-memory errors.

CREATING AN EMPTY FREE STORE HANDLER

As you learned in Lesson 33, when the *new* operator cannot allocate memory from the free store, *new* assigns the NULL value to your pointer variable. The following program *Use_Free.CPP*, repeatedly invokes the *new* operator, allocating 1,000 bytes until the free store is empty:

```
#include <iostream.h>

void main(void)
 {
   char *pointer;

   do {
     pointer = new char[1000];

     if (pointer != NULL)
       cout << "Allocated 1000 bytes" << endl;
     else
       cout << "Free store empty" << endl;

   } while (pointer);
 }
```

As you can see, the *Use_Free.CPP* program loops until the *new* operator assigns the NULL value to the pointer. When you want *new* to perform different processing when it cannot satisfy a memory request (something other than simply returning the NULL value), you first define a function you want your program to invoke when insufficient memory for a request exists. For example, the following function, *end_program*, displays a message on the screen and then uses the *exit* run-time library function to end the program:

```
void end_program(void)
 {
   cout << "Memory request cannot be satisfied" << endl;
   exit(1);
 }
```

To direct C++ to call the *end_program* function when *new* cannot satisfy a memory request, you invoke the *set_new_handler* function with the *end_program* function as a parameter, as shown here:

```
set_new_handler(end_program);
```

The following program, *End_Free.CPP*, invokes the *end_program* function when the *new* operator cannot satisfy a memory request:

```
#include <iostream.h>
#include <stdlib.h>   // Exit prototype
#include <new.h>       // set_new_handler prototype

void end_program(void)
 {
   cout << "Memory request cannot be satisfied" << endl;
   exit(1);
 }

void main(void)
 {
   char *pointer;

   set_new_handler(end_program);

   do {
     pointer = new char[10000];

     cout << "Allocated 10000 bytes" <<endl;

   } while (1);
 }
```

In this example, the program simply ends when the *new* operator cannot allocate memory from the free store. Depending on your program's needs, you could use the function to allocate memory from another source, such as the computer's extended memory, which exists in the MS-DOS environment. In addition, your program could chose to release memory it previously allocated for other purposes, to make free store memory available. By providing your programs with the ability to customize the insufficient-memory handler, C++ gives you complete control over the memory-allocation process.

Note: The **Turbo C++ Lite** compiler this book's companion CD-ROM provides does not support the **set_new_handler** function.

CREATING YOUR OWN NEW AND DELETE OPERATORS

As you have learned, C++ lets your programs overload operators. In this way, you can overload the *new* and *delete* operators to change their behavior. For example, assume that you allocate 100 bytes of memory to hold your company's super-secret data. When you later release the memory using the *delete* operator, the program will release the buffer that stored the data. Assuming that a "hot shot" corporate spy (and C++ programmer) has access to your computer, the programmer could, in theory, write a program that locates the 100-byte array in your computer's memory and displays your super secrets.

By overloading the *delete* operator, your program can first fill the buffer with zeros or other nonsense characters before releasing the memory. The following program, *MyDelete.CPP*, overloads the *delete* operator. The program first overwrites the 100 bytes pointed to by the pointer before releasing the memory, using the *free* run-time library function:

```cpp
#include <iostream.h>
#include <stdlib.h>
#include <string.h>

static void operator delete(void *pointer)
  {
    char *data = (char *) pointer;
    int i;

    for (i = 0; i < 100; i++)
      data[i] = 0;

    cout << "The secrets are safe!" << endl;

    free(pointer);
  }

void main(void)
  {
    char *pointer = new char[100];

    strcpy(pointer, "My Company Secrets");

    delete pointer;
  }
```

When the program begins, it allocates memory for a character-string array using *new*. The program then copies your company's secrets to the string. Later, the program uses the overloaded *delete* operator to discard the memory. Within the *delete* function, the following statement assigns the pointer to a character string pointer:

```cpp
char *data = (char *) pointer;
```

The characters *(char *)* are a *cast* whose purpose is simply to tell the C++ compiler that the function knows it is assigning a pointer of type *void* (see previous function parameter) to a pointer of type *char*. If you eliminate the cast, the program will not compile. Next, the function copies zeros to the buffer's 100 bytes and then releases the memory using the run-time library *free* function. It is important to note that this function only works for a memory region of 100 bytes. Because the *MyDelete.CPP* program only allocates memory one time, the *free* function works and the program executes correctly. If you change the program to only allocate 10 bytes of memory and you do not make the same change to the function, the function will overwrite 90 bytes of memory that your program might be using in some other way, causing errors. Using run-time library functions, however, your programs can learn more information about the size of the memory region a pointer references. For more information on memory and the free store (C programmers call the free store the *heap*), see the book *Jamsa's C/C++ Programmer's Bible*, Jamsa Press, 1997.

The previous program, *MyDelete.CPP*, overloaded the *delete* operator. In a similar way, the following program, *New_Over.CPP*, overloads the C++ *new* operator. In this case, the overloaded operator places the character string "Rescued by C++!" at the start of the allocated memory:

```
#include <iostream.h>
#include <alloc.h>
#include <string.h>

static void *operator new(size_t size)
  {
    char *pointer;

    pointer = (char *) malloc(size);

    if (size > strlen("Rescued by C++!"));
      strcpy(pointer, "Rescued by C++!");

    return(pointer);
  }

void main(void)
  {
    char *str = new char[100];

    cout << str << endl;
  }
```

As you can see, the new function uses the *malloc* run-time library function to allocate the memory. If the size of the allocated memory is large enough to hold the string "Rescued by C++!", the function uses the *strcpy* run-time library function to copy the string to the memory region. By overloading the *new* and *delete* operators, your programs gain considerable control over memory-allocation operations.

WHAT YOU MUST KNOW

As your programs become more complex, they will allocate memory during execution using the *new* operator. In this lesson, you learned how to change the *new* operator's behavior by first defining the handler function your program invokes when *new* cannot satisfy a memory request and then by overloading the *new* operator itself. In Lesson 35, "Getting More from *cin* and *cout*," you will learn new ways to use the *cin* and *cout* input and output streams to improve your program's input and output capabilities. Before you continue with Lesson 35, however, make sure that you have learned the following key concepts:

☑ By default, when the *new* operator cannot satisfy a memory request, *new* will return the NULL value.

☑ If your programs need different processing to occur when *new* cannot satisfy a memory request, your programs can define their own handlers. Using the *set_new_handler* function your program can direct *new* to call your function when it cannot satisfy a memory request.

☑ C++ lets your programs overload the *new* and *delete* operators. However, before you do so, you should have a strong understanding of the free store (the heap) and the run-time library routines that manipulate it.

LESSON 35

GETTING MORE FROM CIN AND COUT

Throughout this book, you have used the *cout* output stream to display information to your screen display. Likewise, many of your programs have used the *cin* input stream to read information from the keyboard. If you examine the header file *iostream.h*, you will find that the file defines *cin* and *cout* as class objects. As objects, *cin* and *cout* support different operators and different operations. In this lesson, you will learn how to increase your program's input and output capabilities using methods built into the *cin* and *cout* classes. By the time you finish this lesson, you will understand the following key concepts:

- The header file *iostream.h* contains class definitions you can examine to better understand I/O streams.

- Using the *cout.width* method, your programs can control the output spacing.

- Using the *cout.fill* method, your programs can replace whitespace output (tabs and spaces) with a specific character.

- To control the number of floating-point digits *cout* displays, your programs can use the *cout.setprecision* method.

- To display output one character at a time, or to perform input one character at a time, your programs can use the *cout.put* and *cin.get* methods.

- Using the *cin.getline* method, your programs can input an entire line at one time.

Almost every C++ program you create will use *cout* or *cin* to perform I/O operations. Take time to experiment with the programs this lesson presents.

TAKING A LOOK INSIDE *IOSTREAM.H*

Since Lesson 2, "Creating Your First Program," every C++ program you have written has included the header file *iostream.h*. If you examine the file's contents, you will find that the file contains definitions that let your programs use *cout* to perform output and *cin* to perform input. More specifically, this file defines the *istream* and *ostream* (input stream and output stream) classes of which *cin* and *cout* are object variables.

Take time now to print a copy of the file *iostream.h*. The file will reside in your compiler's *INCLUDE* subdirectory. The definitions in the file are quite complex. However, if you traverse the file slowly, you will find that most of the definitions are simply class and constant definitions. Within the file you will find the declarations for the objects *cin* and *cout*.

PUTTING COUT TO USE

As you have read, *cout* is a class that contains several different methods. The following programs illustrate how to use several *cout* methods within your programs to format your output. As you learned in Lesson 4, "Writing Messages to Your Screen," the *setw* manipulator lets your programs specify the minimum number of spaces the next value the program outputs will consume, as shown here:

```
#include <iostream.h>
#include <iomanip.h>
```

```
void main(void)
 {
   cout << "My favorite number is" << setw(3) << 1001 << endl;
   cout << "My favorite number is" << setw(4) << 1001 << endl;
   cout << "My favorite number is" << setw(5) << 1001 << endl;
   cout << "My favorite number is" << setw(6) << 1001 << endl;
 }
```

Like the *setw* manipulator, the *cout.width* method lets you specify the minimum number of characters *cout* will use to output the value that follows the function call. The following program, *CoutWidt.CPP*, uses the *cout.width* method to perform processing identical to that of *setw*, just shown:

```
#include <iostream.h>
#include <iomanip.h>

void main(void)
 {
   int i;

   for (i = 3; i < 7; i++)
    {
      cout << "My favorite number is";
      cout.width(i);
      cout << 1001 << endl;
    }
 }
```

When you compile and execute the *CoutWidt.CPP* program, your screen will display the following output:

```
C:\> CoutWidt   <ENTER>
My favorite number is1001
My favorite number is1001
My favorite number is 1001
My favorite number is  1001
```

Like the *setw* manipulator, the width your program selects using the *cout.width* method is only in effect for the next value the program outputs.

USING A CHARACTER PAD

When you use the *setw* manipulator or the *cout.width* method to control output spacing, *cout* will place blanks before values (or after, for left-justified), as required. Depending on your program, there may be times when you want to use a character other than a blank. For example, assume your program creates a table such as the following:

```
Table of Information
Company Profile.........................................10
Company Profit and Loss.................................11
Company Board Members...................................13
```

In this example, the output precedes the page numbers with dots (called dot leaders). The *cout.fill* method lets you specify the character *cout* uses to fill whitespace. The following program, *CoutFill.CPP*, creates a table similar to that just shown:

```
#include <iostream.h>
#include <iomanip.h>

void main(void)
 {
   cout << "Table of Information" << endl;
   cout.fill('.');
   cout << "Company Profile" << setw(20) << 10 << endl;
   cout << "Company Profit and Loss" << setw(12) << 11 << endl;
   cout << "Company Board Members"<< setw(14) << 13 << endl;
 }
```

After you use the *cout.fill* method to select the fill character, your selection will remain in effect until you change it using *cout.fill* a second time.

CONTROLLING FLOATING-POINT DIGITS

By default, when you use *cout* to display a floating-point value, you are never quite sure how many digits *cout* will display. Using the *setprecision* manipulator, however, you can specify the number of digits you desire. The following program, *SetPrec.CPP*, uses the *setprecision* manipulator to control the number of digits that appear to the right of the decimal point:

```
#include <iostream.h>
#include <iomanip.h>

void main(void)
 {
   float value = 1.23456;

   int i;

   for (i = 1; i < 6; i++)
     cout << setprecision(i) << value << endl;
 }
```

When you compile and execute the *SetPrec.CPP* program, your screen will display the following output:

```
C:\> SetPrec   <ENTER>
1.2
1.23
1.235
1.2346
1.23456
```

When you use the *setprecision* manipulator to change the precision, your selection will remain in effect until your program uses *setprecision* a second time.

PERFORMING OUTPUT ONE CHARACTER AT A TIME

Depending on your program, there may be times when you must display characters one at a time or read keyboard input one character at a time. To display output one character at a time, your programs can use the *cout.put* method. The following program, *CoutPut.CPP*, uses this method to display the message *Rescued by C++!* to the screen one character at a time:

```cpp
#include <iostream.h>

void main(void)
  {
    char string[] = "Rescued by C++!";
    int i;

    for (i = 0; string[i]; i++)
      cout.put(string[i]);
  }
```

The run-time library provides a function named *toupper*, which returns the uppercase equivalent of a lowercase letter. The following program *CoutUppr.CPP*, uses the *toupper* function to convert a character to uppercase and then displays the letter using *cout.put*:

```cpp
#include <iostream.h>
#include <ctype.h>        // toupper prototype

void main(void)
  {
    char string[] = "Rescued by C++!";
    int i;

    for (i = 0; string[i]; i++)
      cout.put(toupper(string[i]));

    cout << endl << "Ending string: " << string << endl;
  }
```

When you compile and execute the *CoutUppr.CPP* program, your screen will display the following output:

```
C:\> CoutUppr    <ENTER>
RESCUED BY C++!
Ending string: Rescued by C++!
```

READING KEYBOARD INPUT ONE CHARACTER AT A TIME

Just as *cout* provides the *cout.put* method to display a character, *cin* provides the method *cin.get*, which lets you read one character of data. To use *cin.get*, you simply assign the character the method returns to a variable, as shown here:

```cpp
letter = cin.get();
```

The following program, *Cin_Get.CPP*, displays a message prompting you to enter a Y or N response. The program then repeatedly loops using *cin.get* to read characters until it encounters a Y or N:

```cpp
#include <iostream.h>
#include <ctype.h>

void main(void)
 {
   char letter;

   cout << "Do you want to continue? (Y/N): ";

   do {
     letter = cin.get();
     // Convert to uppercase
     letter = toupper(letter);
   } while ((letter != 'Y') && (letter != 'N'));

   cout << endl << "You selected " << letter << endl;
 }
```

READING KEYBOARD INPUT ONE LINE AT A TIME

As you have learned, when you use *cin* to perform input operations, *cin* uses whitespace characters, such as blank, tab, or carriage return, to determine where one value ends and a second begins. In many cases, you will want your programs to read an entire line of data into a character string. To do so, your programs can use the *cin.getline* method. To use *cin.getline*, you specify a character string into which the method places the letters and you specify a parameter that specifies the string size, as shown here:

```cpp
cin.getline(string, 64);
```

As *cin.get* reads characters from the keyboard, it will not read more characters than the string can hold. A convenient way to express the string size is to use the C++ *sizeof* operator, as shown here:

```cpp
cin.getline(string, sizeof(string));
```

Should you later change the size of your string, you will not have to search for and change each *cin.get* statement that appears in your program. Instead, the *sizeof* operator will use the correct string size. The following program, *GetLine.CPP*, uses the *cin.getline* method to read a line of text from the keyboard:

```cpp
#include <iostream.h>

void main(void)
 {
   char string[128];

   cout << "Type line of text and press Enter" << endl;

   cin.getline(string, sizeof(string));

   cout << "You typed: " << string << endl;
 }
```

When your program reads characters from the keyboard, there may be times when you must read characters up to and including a specific letter. After the program reads the specific letter, you will want the current input operation to end. To perform such an operation, your program can pass the desired letter to the *cin.getline* method. For example, the following function call directs *cin.getline* to read a line of text, ending the input operation when it encounters a carriage return, 64 characters, or the letter Z:

```
cin.getline(string, 64, 'Z');
```

The following program *Until_Z.CPP*, uses the *cin.getline* method to read a line of text or characters up to and including the letter Z:

```
#include <iostream.h>

void main(void)
 {
    char string[128];

    cout << "Type line of text and press Enter" << endl;

    cin.getline(string, sizeof(string), 'Z');

    cout << "You typed: " << string << endl;
 }
```

Compile and execute the *Until_Z.CPP* program. Experiment by entering different lines of text, such as some that start with the letter Z, some that end with the letter Z, and some that do not contain the letter Z at all.

WHAT YOU MUST KNOW

Every C++ program you create will probably use *cin* or *cout* to perform input and output operations. This lesson has introduced several I/O manipulators and functions you can use with the *cin* and *cout* streams. As your programs become more complex, they will often store information in files. In Lesson 36, "C++ File I/O Operations," you will learn how to perform file input and output operations in C++. Before you continue with Lesson 36, however, make sure that you have learned the following key concepts:

- ☑ The header file *iostream.h* defines the *cin* and *cout* streams as object variables of the *istream* and *ostream* classes. As class objects, *cin* and *cout* provide methods your programs can invoke to perform specific tasks.

- ☑ The *cout.width* method lets your programs specify the minimum number of characters the next value output will use.

- ☑ The *cout.fill* method lets your programs specify the character *cout* will use to add spaces for *cout.width* or *setw*.

- ☑ The *setprecision* manipulator lets your programs control the number of digits displayed to the right of the decimal point for floating-point values.

- ☑ The *cin.get* and *cout.put* methods let your programs input or output one character at a time.

- ☑ The *cin.getline* method lets your program read a line of text from the keyboard.

LESSON 36

C++ FILE I/O OPERATIONS

As your programs become more complex, they will store and retrieve information using files. If you are familiar with file manipulation in the C programming language, you can use similar techniques within C++. In addition, as you will learn in this lesson, C++ provides a collection of file stream classes that make file input and output (I/O) operations very easy. By the time you finish this lesson, you will understand the following key concepts:

- Using an output file stream, you can write information to the file using the *insertion operator* (<<).

- Using an input file stream, you can read information stored in a file using the *extraction operator* (>>).

- To open and close files, you use class methods.

- To read and write file data, you can use the insertion and extraction operators as well as several file class methods.

Many of the programs you create in the future will make extensive use of files. Take time to experiment with the programs this lesson presents. As you will find, C++ makes file operations very easy.

WRITING OUTPUT TO A FILE STREAM

In Lesson 35, "Getting More from *cin* and *cout*," you learned that *cout* is an object of type *ostream* (output stream). Using the *ostream* class, your programs can perform output to *cout* using the insertion operator or different class methods, such *cout.put*. The header file *iostream.h* defines the *cout* output stream.

In a similar way, the header file *fstream.h* defines an output file stream class named *ostream*. Using *ofstream* class objects, your programs can perform file output. To begin, you declare an *ofstream* object, specifying the desired filename as a character string, as shown here:

```
ofstream  file_object("FILENAME.EXT");
```

When you specify a filename within a declaration of an *ofstream* object, C++ will create a new file on your disk using the given name. If a file with same name already exists on your disk, C++ will overwrite the file's contents. The following program, *Out_File.CPP*, creates an *ofstream* object and then uses the insertion operator to output several lines of text to the file *BookInfo.DAT*:

```
#include <fstream.h>

void main(void)
 {
   ofstream book_file("BookInfo.DAT");

   book_file << "Rescued by C++, Third Edition" << endl;
   book_file << "Jamsa Press" << endl;
   book_file << "29.95" << endl;
 }
```

237

In this example, the program opens the file *BookInfo.DAT* and then writes three lines to the file using the insertion operator. Compile and execute the *Out_File.CPP* program. If you are working in the MS-DOS environment, you can use the *TYPE* command to display the file's contents, as shown here:

```
C:\> TYPE  BookInfo.DAT   <ENTER>
Rescued by C++, Third Edition
Jamsa Press
$29.95
```

As you can see, C++ makes simple file output operations easy.

READING FROM AN INPUT FILE STREAM

As you just learned, using the *ofstream* class, your programs can quickly perform file output operations. In a similar way, your programs can perform file input operations using *ifstream* objects. To access a file for input operations, you simply create an object and pass the object's constructor method the name of the file you want, as shown here:

```
ifstream input_file("FILENAME.EXT");
```

The following program, *File_In.CPP*, opens the file *BookInfo.DAT*, which you created in the previous program, and reads and displays the file's first three entries:

```
#include <iostream.h>
#include <fstream.h>

void main(void)
  {
    ifstream input_file("BookInfo.DAT");

    char one[64], two[64], three[64];

    input_file >> one;
    input_file >> two;
    input_file >> three;

    cout << one << endl;
    cout << two << endl;
    cout << three << endl;
  }
```

When you compile and execute the *File_In.CPP* program, you might expect the program to display the file's first three lines. However, like *cin*, file input streams use whitespace to indicate where one value ends and another begins. As a result, when you execute the previous program, your screen will display the following output:

```
C:\> File_In   <ENTER>
Rescued
by
C++,
```

READING A COMPLETE LINE OF FILE INPUT

In Lesson 35, "Getting More From *cin* and *cout*" you learned that your program can use *cin.getline* to read an entire line from the keyboard. In a similar way, *ifstream* objects can use *getline* to read a line of file input. The following program, *FileLine.CPP*, uses the *getline* method to read all three lines of the file *BookInfo.DAT*:

```cpp
#include <iostream.h>
#include <fstream.h>

void main(void)
 {
   ifstream input_file("BookInfo.DAT");

   char one[64], two[64], three[64];

   input_file.getline(one, sizeof(one));
   input_file.getline(two, sizeof(two));
   input_file.getline(three, sizeof(three));

   cout << one << endl;
   cout << two << endl;
   cout << three << endl;
 }
```

In this example, the program successfully reads the file's contents because it knew the file contained three lines. In many cases, however, your program will not know how many lines the file contains. In such cases, your programs will simply continue reading the file's contents until they encounter the file's end, as discussed next.

TESTING FOR THE END OF A FILE

Within your programs, a common file operation is to read a file's contents up to the end of the file. To test for the end of a file, your programs can use the object's *eof* method. The method returns the value 0 if it has not yet encountered the end of the file and 1 if the end of file has occurred. Using a *while* statement, programs can loop, reading the file's contents up to the end of the file, as shown here:

```cpp
while (! input_file.eof())
 {
   // Statements
 }
```

In this example, the program will continue to loop as long as the *eof* method returns false (0). The following program, *Test_Eof.CPP*, uses the *eof* method to read the contents of the file *BookInfo.DAT* until it encounters the end of the file:

```cpp
#include <iostream.h>
#include <fstream.h>

void main(void)
 {
```

```
    ifstream input_file("BookInfo.DAT");

    char line[64];

    while (! input_file.eof())
      {
        input_file.getline(line, sizeof(line));

        cout << line << endl;
      }
  }
```

Similar to the *Test_Eof.CPP* program, the following program, *Word_EOF.CPP*, reads the file's contents one word at a time until it encounters the end of the file:

```
#include <iostream.h>
#include <fstream.h>

void main(void)
  {
    ifstream input_file("BookInfo.DAT");

    char word[64];

    while (! input_file.eof())
      {
        input_file >> word;

        cout << word << endl;
      }
  }
```

As you can see, the *Word_EOF.CPP* program uses the extraction operator (>>) to read one word from the input file stream.

Finally, the following program, *Char_EOF.CPP* reads the file's contents one character at a time, using the *get* method until it encounters the end of the file:

```
#include <iostream.h>
#include <fstream.h>

void main(void)
  {
    ifstream input_file("BookInfo.DAT");

    char letter;

    while (! input_file.eof())
      {
        letter = input_file.get();

        cout << letter;
      }
  }
```

TESTING FOR FILE OPERATION ERRORS

So far, this lesson's programs have assumed that no errors occur during file I/O operations. Unfortunately, that is not always the case. For example, when your program opens a file for input, your program should ensure that the file exists. Likewise, when your program writes data to a file, you must ensure that the operation was successful (a *disk-full* error, for example, can prevent the disk from recording data to the file). To help your programs test for errors, you can use a file object's *fail* method. If no error occurs during a file operation, the *fail* method will return false (0). If an error occurs, however, the *fail* method will return true. For example, when your program opens a file, your program should use the *fail* method to determine if an error has occurred, as shown here:

```
ifstream input_file("FILENAME.DAT");

if (input_file.fail())
  {
     cerr << "Error opening FILENAME.EXT" << endl;
     exit(1);
  }
```

In a similar way, your program should test to ensure that its file read and write operations are successful. The following program, *Test_All.CPP*, uses the *fail* method to test different error conditions:

```
#include <iostream.h>
#include <fstream.h>

void main(void)
  {
    char line[256];

    ifstream input_file("BookInfo.DAT");

    if (input_file.fail())
     cerr << "Error opening BookInfo.DAT" << endl;
    else
     {
       while ((! input_file.eof()) && (! input_file.fail()))
        {
           input_file.getline(line, sizeof(line));

          if (! input_file.fail())
             cout << line << endl;
        }
     }
  }
```

CLOSING A FILE WHEN YOU NO LONGER NEED IT

When your program ends, the operating system will close files that the program has open. As a rule, however, when your program no longer needs a file, your program should close it. To close a file, your program should use the *close* method, as shown here:

```
input_file.close();
```

When you close a file, the operating system will flush (record) to disk any data your program has written to the file, and will update the file's directory entry.

CONTROLLING HOW A PROGRAM OPENS A FILE

In each of this lesson's example programs, file input and output operations begin at the start of the file. When your program writes data to an output file, however, there will be times when you may want the program to append information to the end of the file. To open a file in *append mode*, you must specify a second parameter (*ios::app*) when you open the file, as shown here:

```
ifstream output_file("FILENAME.EXT", ios::app);
```

In this case, the parameter *ios::app* specifies a *file-open mode*. As your programs become more complex, they use a combination of the open mode values specified in Table 36.

Open Mode	Purpose
ios::app	Opens a file in append mode, placing the file pointer at the end of the file
ios::ate	Places the file pointer at the end of the file
ios::in	Opens the file for input
ios::nocreate	If the file does not exist, do not create the file and return an error
ios::noreplace	If the file exists, the file-open operation should fail and return an error
ios::out	Opens the file for output
ios::trunc	Truncates (overwrites) the contents of an existing file

Table 36 File-open mode values.

As you have learned, when your programs open a file for output, the file methods normally overwrite an existing file's contents. Depending on your program's purpose, there may be times when you do not want to overwrite an existing file. The following file-open operation, for example, opens a file for output, using the *ios::noreplace* mode to prevent the program from overwriting an existing file:

```
ifstream output_file("FILENAME.EXT", ios::out | ios::noreplace);
```

PERFORMING READ AND WRITE OPERATIONS

So far, this lesson's programs have performed file operations using only character strings. As your programs become more complex, there will be times when you must read and write arrays and structures. In such cases, your programs can use *read* and *write* methods. When you use the *read* method you specify a data buffer into which the method reads the data and a second parameter that specifies the buffer size in bytes, as shown here:

```
input_file.read(buffer, sizeof(buffer));
```

Likewise, when you use the *write* method, you specify both the buffer that contains the data the method is to write to the file, and a parameter that specifies the buffer size, as shown here:

```
output_file.write(buffer, sizeof(buffer));
```

For example, the following program, *Stru_Out.CPP*, uses the *write* method to output the contents of a structure to the file *Employee.DAT*:

```
#include <iostream.h>
#include <fstream.h>

void main(void)
 {
   struct employee {
     char name[64];
     int age;
     float salary;
   } worker = { "John Doe", 33, 25000.0 };

   ofstream emp_file("Employee.DAT");

   emp_file.write((char *) &worker, sizeof(employee));
 }
```

The *write* method normally receives a pointer to a character string. Within the *write* method call, the letters *(char *)* are a *cast*, which informs the compiler that you are passing a pointer of a different type. If you omit the cast, the C++ compiler will generate a syntax error because the function expects a pointer to the type *char*. In a similar way, the following program, *Stru_In.CPP*, uses the *read* method to read the employee information from the file:

```
#include <iostream.h>
#include <fstream.h>

void main(void)
 {
   struct employee {
     char name[64];
     int age;
     float salary;
   } worker = { "John Doe", 33, 25000.0 };

   ifstream emp_file("Employee.DAT");

   emp_file.read((char *) &worker, sizeof(employee));

   cout << worker.name << endl;
   cout << worker.age << endl;
   cout << worker.salary << endl;
 }
```

243

WHAT YOU MUST KNOW

As your programs become more complex you will perform file operations on a regular basis. Take time to experiment with the programs this lesson presents. In Lesson 37, "Inline Functions and Assembly-Language Code," you will learn how to improve your program performance using inline functions. Before you continue with Lesson 37, however, make sure that you have learned the following key concepts:

☑ The header file *fstream.h* defines the *ifstream* and *ofstream* classes, with which your programs can perform file input and output operations.

☑ To open a file for input or output operations, your program declares an *ifstream* or *ofstream* object, passing the name of the file to the object's constructor method.

☑ After your program opens a file for input and output operations, your program can read or write data using the extraction (>>) and insertion (<<) operators.

☑ Your programs can perform character input and output to and from a file using the *get* and *put* methods.

☑ Your programs can read an entire line from a file using the *getline* method.

☑ Most programs will read a file's contents up to the end of the file. Your programs can test for the end of the file using the *eof* method.

☑ When your programs perform file operations, they should test the status of different operations to ensure that the operations are successful. To perform error testing, your programs can use the *fail* method.

☑ If your programs must input or output data, such as a structure or an array, your programs can use the *read* and *write* methods.

☑ When your program has finished using a file, it should close the file using the *close* method.

LESSON 37

INLINE FUNCTIONS AND ASSEMBLY-LANGUAGE CODE

Since Lesson 11, "Getting Started with Functions," your programs and classes have used functions extensively. As you have learned, the only disadvantage of using functions is that they introduce overhead (increase your program's processing time) because they must place parameters onto the stack for each invocation. As you will learn in this lesson, for short functions you can use a technique called *inline code*, which places the equivalent function statements at each function call throughout your program—which, in turn, eliminates the function-call overhead. Using inline functions, your programs will execute slightly faster. By the time you finish this lesson, you will understand the following key concepts:

- To improve performance by reducing function call overhead, you can direct the C++ compiler to place a function's code inline, much like a macro substitution.

- By using *inline functions,* your program remains readable (someone reading your program sees the function call), but you eliminate the function overhead that occurs from pushing and popping values from the stack. You also eliminate the overhead from jumping to and returning from the function.

- Depending on your program's requirements, there are times when you must use assembly language to accomplish a specific task.

- To simplify your use of assembly-language programming, C++ lets you specify inline assembly-language functions within your C++ programs.

UNDERSTANDING INLINE FUNCTIONS

When you define a function within your program, the C++ compiler converts the function code to machine language, keeping only one copy of the instructions within your program. Each time your program calls the function, the C++ compiler places special instructions in your program that place the function's parameters onto the stack and then branch the program's execution to the function instructions. When the function statements end, the program's execution continues at the first statement that follows the function call. The code that places arguments onto the stack and then branches the program's execution to and from the function statements introduces overhead that causes your programs to execute slightly slower than if the program placed the same statements inline within your program at each function reference. For example, consider the following program, *CallBeep.CPP*, which invokes the *show_message* function, which in turn beeps the computer's speaker a specific number of times and then displays a message:

```
#include <iostream.h>

void show_message(int count, char *message)
  {
    int i;

    for (i = 0; i < count; i++)
      cout << '\a';

    cout << message << endl;
  }
```

245

```
void main(void)
  {
    show_message(3, "Rescued by C++");

    show_message(2, "Lesson 37");
  }
```

The following program, *No_Call.CPP*, does not invoke the *show_message* function. Instead, the program places the function's same statements inline at each function reference, as shown here:

```
#include <iostream.h>

void main(void)
  {
    int i;

    for (i = 0; i < 3; i++)
      cout << '\a';

    cout << "Rescued by C++" << endl;

    for (i = 0; i < 2; i++)
      cout << '\a';

    cout << "Lesson 37" << endl;
  }
```

Both programs perform identical processing. Because the *No_Call* program does not invoke the *show_message* function, it executes slightly faster than the *CallBeep* program. In this case, the difference in execution time is impossible to detect; however, if the program invoked the function 1,000 times, you might see a slight performance increase. On the other hand, the *No_Call* program is more cluttered than its *CallBeep* counterpart and might therefore be more difficult for you to understand.

When you create programs, you must always try to determine when using functions is advantageous and when your programs should instead use inline functions. For most simple programs, using functions is usually your best bet. However, if you are writing a program for which performance is paramount, you might need to reduce the number of function invocations your program performs.

One way to reduce function invocations is to place the corresponding statements throughout your program as just done in the *No_Call* program. As you saw, however, eliminating only one function added considerable clutter to your program. Luckily, C++ provides the *inline* keyword that provides the best of both worlds.

USING THE INLINE KEYWORD

When you declare a function within your program, C++ lets you precede the function name with the *inline* keyword. When the C++ compiler encounters the *inline* keyword, the compiler will place the function statements at each function call within the (machine language) executable file. In this way, your C++ program can improve its readability by using functions, while improving performance by eliminating the function-call overhead. The following program, *Inline.CPP*, defines the *max* and *min* functions as *inline*:

```cpp
#include <iostream.h>

inline int max(int a, int b)
 {
   if (a > b)
     return(a);
   else
     return(b);
 }

inline int min(int a, int b)
 {
   if (a < b)
     return(a);
   else
     return(b);
 }

void main(void)
 {
   cout << "The min of 1001 and 2002 is " <<
     min(1001, 2002) << endl;
   cout << "The max of 1001 and 2002 is " <<
     max(1001, 2002) << endl;
 }
```

In this example, the C++ compiler will replace each function call with the function's corresponding statements. The program performance improves without making the program more difficult to understand.

Understanding Inline Functions

 When the C++ compiler encounters the *inline* keyword before a function definition, the compiler will later replace references to the function (calls) with statements equivalent to those the function performs. In this way, your programs improve their execution performance by eliminating the function-call overhead, while still benefiting from the increased readability using functions provides.

INLINE FUNCTIONS AND CLASSES

As you have learned, when you define a class, you define class functions within or outside of the class. For example, the following *employee* class defines its functions within the class itself:

```cpp
class employee {
 public:
   employee(char *name, char *position, float salary)
    {
      strcpy(employee::name, name);
      strcpy(employee::position, position);
```

```
        employee::salary = salary;
      }
    void show_employee(void)
      {
        cout << "Name: " << name << endl;
        cout << "Position: " << position << endl;
        cout << "Salary: $" << salary << endl;
      }
  private:
    char name[64];
    char position[64];
    float salary;
  };
```

Programmers often refer to the process of placing functions within a class as placing the class functions inline. When you create inline class functions in this way, C++ duplicates the function for every object of the class you create, placing the code inline at each method reference. The advantage of such inline code is increased performance. The disadvantage is that your programs can become very large very quickly, depending on your object use. In addition, including the function code within the class definition can clutter a class, making its members difficult for you and other programmers to understand.

To improve the readability of your class definitions, you can pull the class functions from inside of the class definition as you normally do and then precede the function with the *inline* keyword. For example, the following function definition directs the compiler to use inline statements for the *show_employee* function:

```
  inline void employee::show_employee(void)
    {
      cout << "Name: " << name << endl;
      cout << "Position: " << position << endl;
      cout << "Salary: $" << salary << endl;
    }
```

USING ASSEMBLY-LANGUAGE STATEMENTS

As you learned in Lesson 2, "Creating Your First Program," programmers can create programs using a wide variety of programming languages. Then, using a compiler, the programmer can convert the program statements into machine code (the ones and zeros) the computer understands. Each computer type supports an intermediate language, called *assembly language,* which falls in between machine language and a programming language, such as C++.

Assembly language uses different symbols to represent machine-language instructions. Depending on your program's function, there may be times when an advanced program must perform low-level operations that require the use of assembly-language statements. In such cases, you may be able to use the C++ *asm* statement to embed the assembly-language statements within your program. Most programs you create will not require assembly-language statements. The following program, *Use_Asm.CPP,* uses the *asm* statement to insert the assembly-language statements necessary to sound the computer's speaker in the MS-DOS environment:

```
#include <iostream.h>

void main(void)
 {
   cout << "About to sound the bell!" << endl;

   asm {
     MOV AH,2
     MOV DL,7
     INT 21H
   }

   cout << "Done!" << endl;
 }
```

As you can see, using the *asm* statement, the program combines C++ and assembly-language statements.

WHAT YOU MUST KNOW

Inline functions improve your program performance by reducing function overhead. In this lesson you learned how and when to use inline functions within your programs. You also learned that there may be times when your programs must use assembly language to accomplish a specific task. In Lesson 38, "Using Command-Line Arguments," you will learn how your programs can access command-line arguments the user types when he or she runs the program. Before you continue with Lesson 38, however, make sure you have learned the following key concepts:

☑ Placing parameters onto the stack and branching to and from function statements introduces overhead that makes your programs run slower.

☑ The *inline* keyword directs the C++ compiler to replace a function call with statements equivalent to those the function would perform. Because the inline statements eliminate the function-call overhead, the program will execute faster.

☑ When you use inline functions within a class, each object you create uses its own inline statements. Normally, objects of the same class type all share the same function code.

☑ The *asm* keyword lets your programs embed assembly-language statements within your C++ programs.

LESSON 38

USING COMMAND-LINE ARGUMENTS

As you know, when you issue commands at your system prompt, most commands let you include additional information, such as a filename, within your command line. For example, when you use the MS-DOS *COPY* command to copy one file's contents to a second file, you specify both filenames within your command line. Likewise, if your compiler is command-line based, you must include the name of your source file when you invoke your compiler. This lesson examines how your C++ programs access command-line arguments. By the time you finish this lesson, you will understand the following key concepts:

- C++ programs treat command-line arguments as parameters to the *main* function.

- Specifically, C++ passes two (sometimes three) parameters to *main*, which most programs call *argc* and *argv*.

- The *argc* parameter contains a count of the number of command-line arguments the operating system passes to your program.

- The *argv* parameter is an array of pointers to character strings, each of which corresponds to a command-line parameter.

- Depending on your compiler, your programs may have access to a third parameter, *env*, which is an array of character-string pointers to your program's environment entries.

Your program's ability to access command-line arguments will increase the number of ways you can use a single program. For example, you can create your own copy program that you can use to copy any source file you specify as the command-line's first argument to the target file you specify as the second command-line argument. Because the copy program supports command-line arguments, you can use the same program to copy any number of files.

ACCESSING ARGV AND ARGC

When you run a program from a system prompt, the command you type becomes your program's command line, as shown here:

```
C:\> COPY   Source.DOC   Target.DOC   <ENTER>
```

In this example, the command line specifies the command (*COPY*) and two arguments (the filenames *Source.DOC* and *Target.DOC*). To let your programs access the command line, C++ passes two parameters to *main*:

```
void main(int argc, char *argv[])
```

The first parameter, *argc*, contains a count of the number of elements in the *argv* array. Using the previous *COPY* command, for example, the *argc* parameter would contain the value 3 (it includes the command name and two arguments). The following program, *ShowArgc.CPP*, uses the *argc* parameter to display a count of the number of command-line arguments:

```
#include <iostream.h>

void main(int argc, char *argv[])
 {
   cout << "Number of command-line arguments is " << argc << endl;
 }
```

Take time to experiment with the *ShowArgc.CPP* program, invoking it with a different number of parameters, as shown here:

```
C:\> ShowArgc  A  B  C  <ENTER>
Number of command-line arguments is 4
```

Depending on your compiler type, your compiler may consider arguments that are grouped within double quotes as a single argument, as shown here:

```
C:\> ShowArgc   "This is only one argument"   <ENTER>
Number of command-line arguments is 2
```

The second parameter to the *main* function, *argv*, is an array of pointers to character strings that contain the individual command-line entries. Figure 38.1 illustrates, for example, how the elements of the *argv* array would point to the command-line entries.

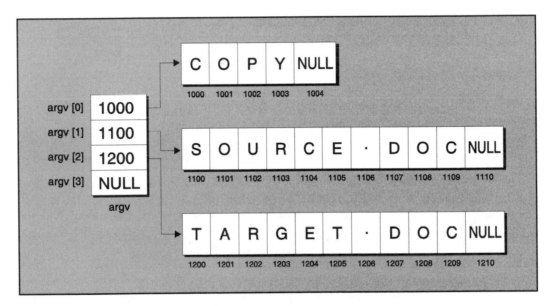

Figure 38.1 *The argv array points to command-line arguments.*

The following program, *ShowArgv.CPP*, uses a *for* statement to display the elements of the *argv* array (the program's command line). The program starts with the array's first element (the program name) and then displays each element until the value of the loop's control variable is greater than *argc* (which contains a count of the number of command-line entries):

```
#include <iostream.h>

void main(int argc, char *argv[])
  {
    int i;

    for (i = 0; i < argc; i++)
      cout << "argv[" << i << "] contains " << argv[i] << endl;
  }
```

After you compile the *ShowArgv.CPP* program, execute it using a command line similar to the following:

```
C:\> ShowArgv A B C   <ENTER>
argv[0] contains ShowArgv.EXE
argv[1] contains A
argv[2] contains B
argv[3] contains C
```

Accessing Command-Line Arguments

To increase the number of tasks a program can perform, C++ lets your program access command-line arguments using two parameters that C++ passes to *main*. The first parameter, *argc*, contains a count of the number of command-line arguments (including the program name). The second parameter, *argv*, is an array of pointers to character strings. Each character string corresponds to a command-line argument. To access command-line arguments, you change the function header for *main*, as shown here:

```
void main(int argc, char *argv[])
```

LOOPING UNTIL ARGV IS NULL

As you have learned, C++ programs use the *NULL* character to terminate character strings. In a similar way, C++ uses a *NULL* character to indicate the last element in the *argv* array. The following program, *ArgvNULL.CPP*, changes the *ShowArgv.CPP* program's *for* statement to loop through the elements of *argv* until *argv's* current element is *NULL*:

```
#include <iostream.h>

void main(int argc, char *argv[])
  {
    int i;

    for (i = 0; argv[i] != NULL; i++)
      cout << "argv[" << i << "] contains " << argv[i] << endl;
  }
```

TREATING ARGV AS A POINTER

As you have learned, C++ lets you access arrays using pointers. The following program, *ArgvPtr.CPP*, treats *argv* as a pointer to a character-string pointer (in other words, a pointer to a pointer) to display the command-line contents:

```
#include <iostream.h>

void main(int argc, char **argv)
  {
    int i = 0;

    while (*argv)
      cout << "argv[" << i++ << "] contains " << *argv++ << endl;
  }
```

Take time to examine the declaration of the *argv* parameter in *main*:

```
void main(int argc, char **argv)
```

The first asterisk in the declaration tells the C++ compiler that *argv* is a pointer. The second asterisk tells the compiler that *argv* is a pointer to a pointer—in this case, a pointer to a pointer of type *char*. Think of *argv* as an array of pointers. Each array element, in this example, points to an array of type *char*.

USING COMMAND-LINE ARGUMENTS

The following program, *FileShow.CPP*, uses command-line arguments to display the contents of a user-specified file to the screen. For example, to use the *FileShow* program to display the contents of the root directory *AUTOEXEC.BAT* file, your command line becomes the following:

```
C:\> FileShow  \AUTOEXEC.BAT  <ENTER>
```

The following statements implement the program, *FileShow.CPP*. The program starts by checking the *argc* parameter to ensure the user has specified a file within the command line. If the user has included a filename, the *argc* parameter will contain the value 2. Next, the program opens and displays the file's contents, using techniques you learned in Lesson 36, "C++ File I/O Operations." As you can see, if the program cannot open the file the user has specified, the program displays an error message and ends:

```
#include <iostream.h>
#include <fstream.h>
#include <stdlib.h>

void main(int argc, char *argv[])
 {
   char line[256];

   if (argc < 2)
    {
      cerr << "You must specify a filename" << endl;
      exit(1);
    }

   ifstream input_file(argv[1]);

   if (input_file.fail())
    cerr << "Error opening BookInfo.DAT" << endl;
   else
    {
      while ((! input_file.eof()) && (! input_file.fail()))
       {
         input_file.getline(line, sizeof(line));

         if (! input_file.fail())
           cout << line << endl;
       }
    }
 }
```

253

ACCESSING OPERATING SYSTEM ENVIRONMENT SETTINGS

As you know, most operating systems let you define environment settings that your programs can access to determine different settings, such as your command path. If you use the MS-DOS environment, for example, you set or display environment settings using the *SET* command. Depending on your compiler type, you may be able to access environment settings within your programs using a third parameter to *main*, named *env*. Like the *argv* parameter, *env* is a pointer to an array of character-string pointers. Also like *argv*, C++ terminates the array with a *NULL* character. If your compiler supports the *env* parameter, you can change your *main* function header, as shown here:

```
void main(int argc, char *argv[], char *env[])
```

The following program, *ShowEnv.CPP*, loops through the elements of the *env* array to display the program's environment settings:

```
#include <iostream.h>

void main(int argc, char *argv[], char *env[])
 {
   while (*env)
     cout << *env++ << endl;
 }
```

As you can see, the program simply loops through the *env* array elements until it encounters the *NULL* pointer, which indicates the array's last entry. When you compile and execute the *ShowEnv.CPP* program, the program will display your environment settings as shown here:

```
C:\> ShowEnv   <ENTER>
TEMP=C:\WINDOWS\TEMP
PROMPT=$p$g
COMSPEC=C:\WINDOWS\COMMAND.COM
PATH=C:\WINDOWS;C:\DOS
```

Accessing Environment Settings

Depending on your compiler type, your programs may be able to access the operating system environment settings using a third command-line parameter, *env*. Like the *argv* parameter, *env* is an array of pointers to character strings, each of which points to an environment entry. To access the environment entries using the *env* parameter, change the function header for main as follows:

```
void main(int argc, char *argv[], char *env[])
```

WHAT YOU MUST KNOW

To increase the number of applications your program can support, C++ lets your programs take advantage of command-line arguments. In Lesson 39, "Understanding and Using Polymorphism," you will examine polymorphism, which lets an object change forms as your program runs. Before you continue with Lesson 39, however, make sure you have learned the following key concepts:

☑ When you run a program from a system prompt, the information you type becomes the program's command line.

☑ To let your programs access the command line, C++ passes two parameters, *argc* and *argv*, to the *main* function.

☑ The *argc* parameter contains a count of the number of command-line arguments.

☑ The *argv* parameter is an array of character-string pointers, each of which contains a command-line argument.

☑ Depending on your compiler, your programs may be able to access a third parameter to *main*, *env*, which is an array of character-string pointers that contain the environment settings.

LESSON 39

UNDERSTANDING AND USING POLYMORPHISM

When programmers talk about C++ and object-oriented programming, one of the terms they will often "toss about" (use but not necessarily understand) is polymorphism. In general, *polymorphism* is an object's ability to change forms. If you break apart the term, you will find that *poly* means many and *morphism* refers to changing forms. A polymorphic object, therefore, is an object that can take on many different forms. This lesson introduces polymorphism and how you use polymorphic objects within your programs to simplify and reduce code. By the time you finish this lesson, you will understand the following key concepts:

- Polymorphism is an object's ability to change forms as your program runs.

- C++ makes it easy for you to create polymorphic objects.

- To create polymorphic objects, your programs must use *virtual* functions.

- A *virtual* function is a base-class function whose name you precede with the keyword *virtual*.

- Any class derived from a base class can use or overload a *virtual* function.

- To create a polymorphic object, you use a pointer to a base-class object.

UNDERSTANDING POLYMORPHISM

A polymorphic object is an object that can change forms as your program executes. For example, assume that you are a programmer who works for the phone company and that you must write a program that simulates phone operations. As you consider ways people use telephones, you quickly identify common operations, such as dialing, ringing, disconnecting, and indicating a busy line. Using these operations, you define the following class:

```cpp
class phone {
  public:
    void dial(char *number) { cout << "Dialing " << number << endl; }
    void answer(void) { cout << "Waiting to answer call" << endl; }
    void hangup(void) { cout << "Done with call—hanging up" << endl; }
    void ring(void) { cout << "Ring, ring, ring" << endl; }
    phone(char *number) { strcpy(phone::number, number); };
  private:
    char number[13];
};
```

The following program, *PhoneOne.CPP*, uses the *phone* class to create a phone object:

```cpp
#include <iostream.h>
#include <string.h>

class phone {
  public:
    void dial(char *number) { cout << "Dialing " << number << endl; }
```

```
    void answer(void) { cout << "Waiting to answer call" << endl; }
    void hangup(void) { cout << "Done with call—hanging up" << endl; }
    void ring(void) { cout << "Ring, ring, ring" << endl; }
    phone(char *number) { strcpy(phone::number, number); };
  private:
    char number[13];
};


void main(void)
{
  phone telephone("555-1212");
  telephone.dial("212-555-1212");
}
```

As you demonstrate your program to your boss, he or she tells you that your program does not differentiate between rotary and touch-tone phones and that it does not support pay-phones, which may require the user to deposit 25 cents before placing a call.

Because you understand inheritance, you decide to derive the *touch_tone* and *pay_phone* classes from the *phone* class as shown here:

```
class touch_tone : phone {
  public:
    void dial(char *number) { cout << "Beep Beep Dialing " << number <<
endl; }
    touch_tone(char *number) : phone(number) { }
};


class pay_phone : phone {
  public:
    void dial(char *number) { cout << "Please deposit " << amount <<
                                " cents" << endl;
                      cout << "Dialing " << number << endl; }
    pay_phone(char *number, int amount) : phone(number) {
                      pay_phone::amount = amount; }
  private:
    int amount;
};
```

As you can see, the *touch_tone* and *pay_phone* classes define their own *dial* method. If you assume the program bases the *phone* class *dial* method on a rotary-dial phone, you do not need to create a rotary-dial phone class. The following program, *NewPhone.CPP*, uses these classes to create *rotary*, *touch_tone*, and *pay_phone* objects:

```
#include <iostream.h>
#include <string.h>

class phone {
  public:
    void dial(char *number) { cout << "Dialing " << number << endl; }
    void answer(void) { cout << "Waiting to answer call" << endl; }
    void hangup(void) { cout << "Done with call—hanging up" << endl; }
    void ring(void) { cout << "Ring, ring, ring" << endl; }
```

```
      phone(char *number) { strcpy(phone::number, number); };
   protected:
      char number[13];
 };

class touch_tone : phone {
  public:
     void dial(char *number){ cout << "Beep Beep Dialing " << number << endl;}
     touch_tone(char *number) : phone(number) { }
 };

class pay_phone : phone {
  public:
     void dial(char *number) { cout << "Please deposit " << amount << " cents"
                                    << endl;
                               cout << "Dialing " << number << endl; }
     pay_phone(char *number, int amount) : phone(number) {
                                    pay_phone::amount = amount; }
  private:
     int amount;
 };

void main(void)
 {
    phone rotary("303-555-1212");
    rotary.dial("602-555-1212");

    touch_tone telephone("555-1212");
    telephone.dial("212-555-1212");

    pay_phone city_phone("555-1111", 25);
    city_phone.dial("212-555-1212");
 }
```

(handwritten annotations: "CLASS" pointing to `phone`, "OBJECT" pointing to `rotary`)

When you compile and execute the *NewPhone.CPP* program, your screen will display the following output:

```
C:\> NewPhone   <Enter>
Dialing 602-555-1212
Beep Beep Dialing 212-555-1212
Please deposit 25 cents
Dialing 212-555-1212
```

As discussed, a polymorphic object is an object that changes forms as your program executes. The previous program, for example, did not use polymorphic objects. In other words, no object changed forms.

MAKING A POLYMORPHIC PHONE OBJECT

After you show your supervisor you new phone program, your supervisor tells you that your phone object must be able to simulate a rotary, touch-tone, or pay-phone on demand. In other words, for one call your phone object may be a touch-tone phone, on the next call a pay-phone, and so on. Therefore, from one phone call to the next your phone object may change forms.

Within the different phone classes, the only function that differs is the *dial* method. To create a polymorphic object, you first make the base-class functions that differ from the derived-class functions *virtual* functions by preceding the function prototype and header with the *virtual* keyword, as shown here:

```cpp
class phone {
  public:
    virtual void dial(char *number) { cout << "Dialing " << number << endl; }
    void answer(void) { cout << "Waiting to answer call" << endl; }
    void hangup(void) { cout << "Done with call—hanging up" << endl; }
    void ring(void) { cout << "Ring, ring, ring" << endl; }
    phone(char *number) { strcpy(phone::number, number); };
  protected:
    char number[13];
};
```

Next, within your program you create a pointer to a base-class object. For your phone program, you will create a pointer to the base-class *phone,* as shown here:

```cpp
phone *poly_phone;
```
   ⌐ CLASS ⌐ POINTER

To change an object's form, you simply assign the address of a derived-class object to the pointer, as shown here:

```cpp
poly_phone = (phone *) &home_phone;
```

The *(phone *)* that follows the assignment operator is a *cast,* which tells the C++ compiler it is OK to assign the address of a variable of one type (*touch_tone*) to a pointer to a variable of a different type (*phone*). Because your program can assign the addresses of different objects to the object-pointer *poly_phone,* the object can change forms and hence, is polymorphic. The following program, *PolyMorp.CPP,* uses these techniques to create a polymorphic phone object. As the program executes, the object *poly_phone* changes from a rotary phone, to a touch-tone phone, to a pay-phone:

```cpp
#include <iostream.h>
#include <string.h>

class phone {
  public:
    virtual void dial(char *number) { cout << "Dialing " << number << endl; }
    void answer(void) { cout << "Waiting to answer call" << endl; }
    void hangup(void) { cout << "Done with call—hanging up" << endl; }
    void ring(void) { cout << "Ring, ring, ring" << endl; }
    phone(char *number) { strcpy(phone::number, number); };
  protected:
    char number[13];
};

class touch_tone : phone {
  public:
    void dial(char *number){ cout << "Beep Beep Dialing " << number << endl;}
    touch_tone(char *number) : phone(number) { }
};
```

259

```
class pay_phone : phone {
  public:
    void dial(char *number) { cout << "Please deposit " << amount <<
                              " cents" << endl;
                              cout << "Dialing " << number << endl; }
    pay_phone(char *number, int amount) : phone(number) {
                              pay_phone::amount = amount; }
  private:
    int amount;
};


void main(void)
 {
   pay_phone city_phone("702-555-1212", 25);
   touch_tone home_phone("555-1212");
   phone rotary("201-555-1212");

   // Make the object a rotary phone
   phone *poly_phone = &rotary;

   poly_phone->dial("818-555-1212");

   // Change object's form to a touchtone phone
   poly_phone = (phone *) &home_phone;
   poly_phone->dial("303-555-1212");

   // Change the object's form to a payphone
   poly_phone = (phone *) &city_phone;
   poly_phone->dial("212-555-1212");
 }
```

When you compile and execute the *PolyMorp.CPP* program, your screen will display the following output:

```
C:\> PolyMorp  <ENTER>
Dialing 818-555-1212
Beep Beep Dialing 303-555-1212
Please deposit 25 cents
Dialing 212-555-1212
```

Because the *poly_phone* object changes forms throughout the program's execution, the object is polymorphic.

Polymorphic Objects Can Change Forms as the Program Executes

A polymorphic object is an object that can change forms as your program executes. To create a polymorphic object, your program uses a pointer to a base-class object. Next, your program assigns the address of a derived-class object to the pointer. Each time the program assigns a different class-type object to the pointer, the pointer object (which is polymorphic) changes forms. Programs build polymorphic objects around base-class *virtual* functions.

UNDERSTANDING PURE VIRTUAL FUNCTIONS

As you have learned, to create a polymorphic object your programs define one or more base-class methods as *virtual* functions. As you derive other classes, the classes can provide their own function that executes in place of the base-class *virtual* function, or they can use the function (in other words, they do not define their own method). Depending on your program, there may be times when it does not make sense for the base class to define the *virtual* function. For example, the derived-object types may differ so greatly that none of them will use the base-class method. In such cases, rather than defining statements for the base-class *virtual* function, your programs can create a *pure virtual function* that has no statements.

To create a pure *virtual* function, your programs specify the function prototype, but they do not specify function statements. Instead, your programs assign the function the value zero, as shown here:

```
class phone {
  public:
    virtual void dial(char *number) = 0;  // Pure virtual function
    void answer(void) { cout << "Waiting to answer call" << endl; }
    void hangup(void) { cout << "Done with call—hanging up" << endl; }
    void ring(void) { cout << "Ring, ring, ring" << endl; }
    phone(char *number) { strcpy(phone::number, number); };
  protected:
    char number[13];
};
```

Each derived class your program defines must define a function for each base-class pure *virtual* function. If a derived class omits a function definition for a pure *virtual* function, the C++ compiler will generate syntax errors.

WHAT YOU MUST KNOW

Polymorphism is an object's ability to change forms as a program executes. This lesson examined the steps you must perform to create polymorphic objects. In Lesson 40, "Using C++ Exceptions to Handle Errors," you will learn how to use C++ exceptions to improve your program's reliability. Before you continue with Lesson 40, however, make sure you have learned the following key concepts:

☑ A polymorphic object can change forms as your program executes.

☑ You create polymorphic objects using classes that you derive from an existing base class.

☑ Within the base class for a polymorphic object, you define one or more functions as *virtual* functions.

☑ In general, polymorphic objects differ by their use of base-class *virtual* functions.

☑ To create a polymorphic object, you create a pointer to a base-class object.

☑ To change a polymorphic object's form, you simply point the object to a different object by assigning the new object's address to the polymorphic object pointer.

☑ A pure *virtual* function is a base-class *virtual* function for which the base class does not define function statements. Instead, the base class assigns the function the value 0.

☑ Derived classes must provide a function definition for each base-class pure *virtual* function.

LESSON 40

USING C++ EXCEPTIONS TO HANDLE ERRORS

After you create and debug (remove the errors from) many programs, you can begin to anticipate the run-time errors the program may encounter. For example, if your program reads information from a file, the program must test whether the file exists and whether or not the program can open it. Likewise, if your program uses the *new* operator to allocate memory, your program must test for and respond to an out-of-memory condition. As your programs increase in size and complexity, you will find that you include many such tests throughout your program. In this lesson, you will learn how to use C++ exceptions to simplify your program's error testing and handling. By the time you finish this lesson, you will understand the following key concepts:

- An *exception* is an unexpected event, an error, within your program.

- Within your program, you define exceptions as classes.

- To direct your programs to watch for exceptions, you use the C++ *try* statement.

- To detect a specific exception, your programs use the C++ *catch* statement.

- To generate an exception when an error occurs, your programs use the C++ *throw* statement.

- When your program detects (catches) an exception, your program calls a special (exception-specific) function called an *exception handler*.

- Some (older) compilers do not support C++ exceptions.

*Note: The **Turbo C++ Lite** compiler provided on this book's companion CD-ROM does not support C++ exceptions.*

C++ REPRESENTS EXCEPTIONS AS CLASSES

Your goal in using C++ exceptions is to simplify error detection and error handling within your programs. Ideally, when your programs experience an unexpected error (an exception), your programs can handle the error in a meaningful way—as opposed to simply ending. Within your program, you define each exception as a class. For example, the following statements define three file-related exceptions:

```
class file_open_error {};
class file_read_error {};
class file_write_error {};
```

Later in this lesson, you will create exceptions that use member variables and functions. For now, however, simply understand that each exception corresponds to a class.

DIRECTING C++ TO TEST FOR EXCEPTIONS

Before your programs can detect and respond to an exception, you must use the C++ *try* statement to enable exception detection. The following *try* statement, for example, enables exception detection for the *file_copy* function call:

```
try {
    file_copy("SOURCE.TXT", "TARGET.TXT");
};
```

Immediately following a *try* statement, your program should place one or more C++ *catch* statements to determine which, if any, exception occurred, as shown here:

```
try {
    file_copy("SOURCE.TXT", "TARGET.TXT");
};


catch (file_open_error) {
  cerr << "Error opening the source or target file" << endl;
  exit(1);
}


catch (file_read_error) {
  cerr << "Error reading the source file" << endl;
  exit(1);
}


catch (file_write_error) {
  cerr << "Error writing to target file" << endl;
  exit(1);
}
```

As you can see, the code tests for the file-exception errors previously defined. In this example, regardless of the error type, the code simply displays a message and exits. Ideally, your code may respond differently, possibly by trying to eliminate the cause of the error so it can retry the operation. If the function is successful and does not generate an exception, C++ will ignore the *catch* statements.

USING THE THROW STATEMENT TO GENERATE AN EXCEPTION

C++ itself does not generate exceptions. Instead, your programs generate an exception using the C++ *throw* statement. For example, within the *file_copy* function, the program can test for and throw an exception, as shown here:

```
void file_copy(char *source, char *target)
  {
    char line[256];

    ifstream input_file(source);
    ofstream output_file(target);

    if (input_file.fail())
     throw(file_open_error);
    else if (output_file.fail())
     throw(file_open_error);
    else
      {
        while ((! input_file.eof()) && (! input_file.fail()))
          {
            input_file.getline(line, sizeof(line));
```

```
        if (! input_file.fail())
          output_file << line << endl;
        else
          throw(file_read_error);
        if (output_file.fail())
          throw(file_write_error);
      }
    }
  }
```

As you can see, the program uses the *throw* statement to generate specific exceptions.

How Exceptions Work

When you use exceptions, your programs test for errors and, if necessary, generate an exception using the *throw* statement. When C++ encounters a *throw* statement, C++ activates the corresponding exception handler (a function whose statements you define within the exception class). After the exception-handling function ends, C++ returns control to the first statement which follows the *try* statement that enabled exception detection. Next, using *catch* statements, your program can determine which exception occurred and can respond accordingly.

DEFINING AN EXCEPTION HANDLER

When your programs throw an exception, C++ executes an exception handler (a function) whose statements you define within the exception class. For example, the following exception class, *nuke_meltdown*, defines exception-handling statements within the *nuke_meltdown* function:

```
class nuke_meltdown {
  public:
    nuke_meltdown(void) { cerr << "\a\a\aRun! Run! Run!" << endl; }
};
```

In this example, when the program throws the *nuke_meltdown* exception, C++ will execute the statements within the *nuke_meltdown* function before it returns control to the first statement which follows the *try* statement that enabled exception detection. The following program, *MeltDown.CPP*, illustrates how to use the *nuke_meltdown* function. The program uses a *try* statement to enable exception detection. Next, the program invokes the *add_u232* function with a parameter. If the parameter's value is less than 255, the function will succeed. If, however, the parameter's value exceeds 255, the function will throw the *nuke_meltdown* exception:

```
#include   <iostream.h>

class nuke_meltdown {
  public:
    nuke_meltdown(void) { cerr << "\a\a\aRun! Run! Run!" << endl; }
};

void add_u232(int amount)
  {
```

```
    if (amount < 255)
      cout << "Amount of u232 is OK" << endl;
    else
      throw nuke_meltdown();
 }


 void main(void)
 {
   try {
      add_u232(255);
   }

   catch (nuke_meltdown) {
     cerr << "Run Faster" << endl;
   }
 }
```

When you compile and execute the *MeltDown.CPP* program, your screen will display the following output:

```
C:\> MeltDown   <ENTER>
Run! Run! Run!
Run Faster
```

If you track the program source code that generates each message, you can follow the exception's flow through the exception handler and back to the *catch* statement. For example, the exception handler, the *nuke_meltdown* function, generates the first line of output. Then, the *catch* statement that detects the exception generates the second line of output.

Defining an Exception Handler

When C++ detects an exception within your program, C++ executes a special function called the exception handler. To define an exception handler, you simply create a function within the exception class that has the same name as the exception (similar to a constructor). When your program later throws an exception, C++ will automatically invoke your exception handler. Ideally, the exception handler should perform operations that remedy the error so your program can retry the operation that caused the exception. After your exception handler ends, your program's execution will continue at the first statement which follows the *try* statement that enabled exception detection.

USING EXCEPTION MEMBER VARIABLES

In the previous examples, your programs could determine, using a *catch* statement, which exception occurred and respond accordingly. Ideally, the more information your programs can obtain about an exception, the better your programs can respond to the error. For example, in the case of a *file_open_error* exception, your program must know the name of the file that caused the error. Likewise, for a *file_read_error* or *file_write_error* exception, the program may want to know the byte location at which the error occurred. To store such information about an exception, your programs simply add member variables to the exception class. When your program later throws an exception, your program passes the information to the exception handling function as parameters, as shown here:

```
throw file_open_error(source);
throw file_read_error(344);
```

Within the exception handler, the function includes statements that assign the parameters to the corresponding class member variables (much like a constructor function). For example, the following statements change the *file_open_error* exception to assign the name of the file that caused the error to the corresponding member variable:

```
class file_open_error {
   public:
    file_open_error(char *filename) {
        strcpy(file_open_error::filename,filename); }
    char filename[255];
};
```

HANDLING UNEXPECTED EXCEPTIONS

In Lesson 13, "Taking Advantage of the Run-Time Library," you learned that C++ compilers provide a run-time library of functions you can use within your programs. As you read the documentation for these functions, you may encounter functions that throw specific exceptions. In such cases, your programs should test for the corresponding exceptions. By default, if your program throws an exception that the program does not catch (your program does not have a corresponding exception handler), your program will execute a default handler that C++ provides. In most cases, the default handler will end your program. The following program, *UnCaught.CPP*, illustrates how the default exception handler ends your program's execution:

```
#include <iostream.h>

class some_exception { };

void main(void)
 {
   cout << "About to throw exception" << endl;
   throw some_exception();
   cout << "Exception thrown" << endl;
 }
```

In this example, when the program throws the exception (which the program does not catch), C++ invokes the default exception handler, which terminates the program. Therefore, the program's last statement, which displays the message about the thrown exception, never executes. Rather than using the C++ default exception handler, your programs can define their own default handler. To tell C++ about the program's default handler, your programs use the *set_unexpected* run-time library function. The header file *except.h* defines the *set_unexpected* function prototype.

STATING WHICH EXCEPTIONS A FUNCTION CAN THROW

As you have learned, a function prototype lets you define a function's return and parameter types. When you use exceptions in your programs, you can also use the function prototype to specify the exceptions a function can throw. For example, the following function prototype tells the compiler the *power_plant* function can throw the *melt_down* and *radiation_leak* exceptions:

```
void power_plant(long power_needed) throw (melt_down, radiation_leak);
```

By including the possible exceptions within the function prototype in this way, other programmers who read your code can readily tell for which exceptions they must test when they use the function.

EXCEPTIONS AND CLASSES

When you create classes, there may be times when you want to define exceptions that are specific to the class. To create a class-specific exception, simply include the exception as one of the class *public* members. For example, the following *string_class* definition defines two exceptions:

```
class string {
  public:
    string(char *str);
    void fill_string(*str);
    void show_string(void);
    int string_length(void);
    class string_empty { };
    class string_overflow {};
  private:
    int length;
    char string[255];
};
```

As you can see, the *string* class defines the *string_empty* and *string_overflow* exceptions. Within your program, you can test for a class exception using the global resolution operator and class name, as shown here:

```
try {
  some_string.fill_string(some_long_string);
};

catch (string::string_overflow) {
    cerr << "String length exceeded—characters truncated" << endl;
}
```

WHAT YOU MUST KNOW

Exceptions exist to simplify and enhance your program's error-detection and error-handling capabilities. To test for and detect exceptions, your programs use the *try, catch,* and *throw* statements. Your knowledge of exceptions will round out your C++ programming expertise. Before you continue your C++ programming, however, make sure you have learned the following key concepts:

☑ An exception is an unexpected error within your program.

☑ Your programs should detect and respond to (handle) exceptions.

☑ Within your programs, you define each exception as a class.

☑ You use the *try* statement to direct the C++ compiler to enable exception detection.

☑ You should place a *catch* statement immediately after your *try* statement to determine which, if any, exception occurred.

☑ C++ does not generate exceptions itself. Rather, your programs generate an exception using the *throw* statement.

☑ When your program catches an exception, your program calls a special function called an exception handler.

☑ When your programs use exceptions, you can specify within a function prototype the exceptions a function can throw.

☑ As you use run-time library functions, be aware that some functions may generate exceptions.

☑ If your program throws an uncaught exception, C++ will invoke a default exception handler.

☑ The header file *except.h* specifies prototypes for functions your programs can use to define their own default-terminate and uncaught-exception handlers.

THE C++ PROGRAMMING LANGUAGE

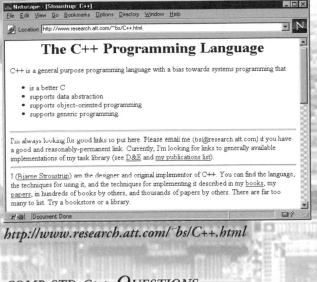

The C++ Programming Language

C++ is a general purpose programming language with a bias towards systems programming that

- is a better C
- supports data abstraction
- supports object-oriented programming
- supports generic programming.

I'm always looking for good links to put here. Please email me (bs@research.att.com) if you have a good and reasonably-permanent link. Currently, I'm looking for links to generally available implementations of my task library (see D&E and my publications list).

I (Bjarne Stroustrup) am the designer and original implementor of C++. You can find the language, the techniques for using it, and the techniques for implementing it described in my books, my papers, in hundreds of books by others, and thousands of papers by others. There are far too many to list. Try a bookstore or a library.

http://www.research.att.com/~bs/C++.html

INFORMATION ABOUT IOSTREAMS

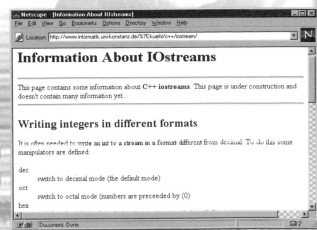

Information About IOstreams

This page contains some information about C++ **iostreams**. This page is under construction and doesn't contain many information yet...

Writing integers in different formats

It is often needed to write an int to a stream in a format different from decimal. To do this some manipulators are defined:

dec
    switch to decimal mode (the default mode)

oct
    switch to octal mode (numbers are preceeded by (0)

hex

http://www.informatik.uni-konstanz.de/%7Ekuehl/—
c++/iostream/

COMP.STD.C++ QUESTIONS

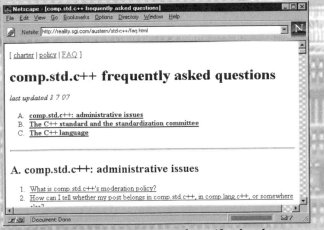

[charter | policy | FAQ]

comp.std.c++ frequently asked questions

last updated 3 7 97

A. comp.std.c++: administrative issues
B. The C++ standard and the standardization committee
C. The C++ language

A. comp.std.c++: administrative issues

1. What is comp.std.c++'s moderation policy?
2. How can I tell whether my post belongs in comp.std.c++, in comp.lang.c++, or somewhere else?

http://reality.sgi.com/austern/std-c++/faq.html

C++ DRAFT STANDARD

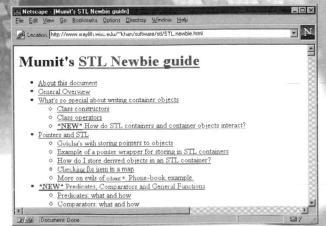

The C++ Draft Standard

The C++ Standard Committee has made the April 28, 1995 revision of the Draft C++ Standard available for public review. There is a home site at Cygnus for the HTML version. *Note that this is a large and formal document, for the use of experts not beginners.* Also, the C++ committee is always working and there are more recent Working Papers that are not available to the public.

FTP Sites

The PostScript and PDF versions can be gotten from:

- AT&T (US)
- University of Warwick (UK)
- University of Sydney (Australia)

The HTML and ASCII conversions can be gotten from ftp.cygnus.com.

http://www-leland.stanford.edu/~iburrell/cpp/std.html

C++/OBJECT ORIENTED LINKS

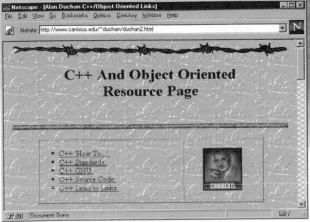

C++ And Object Oriented Resource Page

- C++ 'How To...'
- C++ Standards
- C++ GNU
- C++ Source Code
- C++ Links to Links

http://www.canisius.edu/~duchan/duchan2.html

MUMIT'S STL NEWBIE GUIDE

Mumit's STL Newbie guide

- About this document
- General Overview
- What's so special about writing container objects
  - Class constructors
  - Class operators
  - *NEW* How do STL containers and container objects interact?
- Pointers and STL
  - Gotcha's with storing pointers to objects
  - Example of a pointer wrapper for storing in STL containers
  - How do I store derived objects in an STL container?
  - Checking for item in a map
  - More on evils of char*: Phone-book example.
- *NEW* Predicates, Comparators and General Functions
  - Predicates: what and how
  - Comparators: what and how

http://www.xraylith.wisc.edu/~khan/software/stl/—
STL.newbie.html

G++ FREQUENTLY ASKED QUESTIONS

Netscape - [Frequently asked questions about the GNU C++ compiler - Table of Contents]
File Edit View Go Bookmarks Options Directory Window Help

Location: http://www.cygnus.com/misc/g++FAQ_toc.html

G++ FAQ

Frequently asked questions about the GNU C++ compiler

September 14, 1997

Joe Buck

- The basics: what is g++?
  - What is the latest version of gcc, g++, and libg++?
  - How do I get a copy of g++ for Unix?
  - Getting gcc/g++ for the HP Precision Architecture
  - Getting gcc/g++ binaries for Solaris 2.x
  - How do I get a copy of g++ for (some other platform)?
  - But I can only find g++-1.42!
- Installation Issues and Problems

Document: Done

http://www.cygnus.com/misc/g++FAQ_toc.html

CALLBACKS IN C++

Netscape - [Callbacks in C++]
File Edit View Go Bookmarks Options Directory Window Help

Location: http://www.primenet.com/~jakubik/callback.html

Callbacks in C++

Information on a callback library, a callback paper, new patterns used by the paper, and complete source code for each step presented in the paper.

Free C++ Callback Library - Callback 1.0

A free library based on the final solution described in Callback Implementations in C++. This library allows callbacks to functions (function pointers), function objects, member functions, and member data that are function pointers or function objects. The library has a very permissive copyright to encourage its use.

- callback 1.0: directory with ~15 files totalling ~40 KB. Must use included perl script to generate C++ code.
- callback 1.0.noperl: directory with ~60 files totalling ~130 KB. Library preinstantiated for Callback0, Callback1, Callback2, Callback0wRet, Callback1wRet, and Callback2wRet. No perl script.

Document: Done

http://www.primenet.com/~jakubik/callback.html

HEADER FILE FOR AN ARRAY CLASS

Netscape - [Header File for an Array Class]
File Edit View Go Bookmarks Options Directory Window Help

Location: http://www.informatik.uni-konstanz.de/~kuehl/c++/array.h.html

Header File for an Array Class

Note:

This file is currently under construction. The documentation is not yet complete and the class is not well tested (actually there are only some rudimentary tests...). In addition there are some access methods using iterators missing: This class should be extended to be a useful container for STL algorithms. There are the first steps taken now: a simple iterator support is provided and the first steps towards exception safety are taken. However, the class is still far from being complete...

Table of Contents

This page grew a little too big to be handy without a table of contents. Thus here is a list of items

Netscape

http://www.informatik.uni-konstanz.de/~kuehl/c++/—array.h.html

DATABASE OF PROGRAMMING RESOURCES

Netscape - [DOPER -- Database of Programming Resources]
File Edit View Go Bookmarks Options Directory Window Help

Netsite: http://www-ug.eecg.toronto.edu/~sokorac/Doper/VisualCPP.html

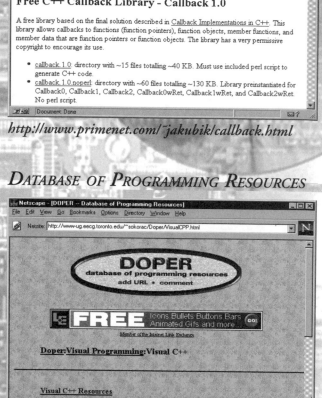

DOPER
database of programming resources
add URL • comment

FREE Icons, Bullets, Buttons, Bars Animated Gifs and more... GO!
Member of the Internet Link Exchange

Doper:Visual Programming:Visual C++

Visual C++ Resources

Document: Done

http://www-ug.eecg.toronto.edu/~sokorac/Doper/—VisualCPP.html

C++ CLASSES

Netscape - [Some C++ classes]
File Edit View Go Bookmarks Options Directory Window Help

Location: http://www.ts.umu.se/~maxell/C++/

C++ *classes*

By Magnus Leksell. C++ FAQ

You might wanna read the introduction!.

- ***Double linked list***
  - dllist.hxx
  - dllist.cxx
  - example
  - polymorphic list example *(not included in the LIB)*
  - String list example *(not included in the LIB)*
  - tlist.hxx - A template class that is based on DLList *(not included in the LIB)*
  - tmpl_list.cxx - An exmple to the template TList class *(not included in the LIB)*

Document: Done

http://www.ts.umu.se/~maxell/C++/

VISUAL C++ RESOURCES

Netscape - [Visual C++ Resources]
File Edit View Go Bookmarks Options Directory Window Help

Netsite: http://www.aul.fiu.edu/tech/visualc.html

AUL **Visual C++** Tech Center

News	Resources on the Web	FAQ's
Visual C++ 4.1 Released	**Microsoft's For Developers Only**	C Frequently asked Questions
Visual C++ Standard Edition announced	**MS's Visual C++ Page**	MFC Frequently asked Questions
VC++ Service Pack 2 available	**Internet Resources for Windows Developers**	
	Win32 Development Knowledge Base	Newsgroups
	Visual C++ Developers Journal	• comp.lang.c
		• comp.lang.c++
		• comp.std.c

Document: Done

http://www.aul.liu.edu/tech/visualc.html

Index

Jamsa Press Runaway Bestsellers

Rescued by Upgrading Your PC, Second Edition

Every day thousands of users experience hardware problems. Worse yet, most users are afraid to open their systems. With the release of *Rescued by Upgrading Your PC, Second Edition*, users should fear no more. This book provides users with easy step-by-step instructions. Not only will users learn how to replace obsolete hardware, they will also learn how to locate system bottlenecks and how to fine-tune system settings.

$24.95 USA
Available: Now!

ISBN: 1-884133-24-X
272 full-color pages • 8 1/2 x 10 3/8"

By Kris Jamsa, Ph.D.

> *Rescued by Upgrading Your PC, Second Edition is a book for the PC users who want to keep their computers on the cutting edge without having to spend a lot of money.*
> —*Northwest View*

Rescued by Personal Computers

To complement our best-selling *Rescued by Upgrading Your PC*, Jamsa Press is pleased to announce *Rescued by Personal Computers*, an introduction to all aspects of computing. This book makes extensive use of full-color illustrations and an easy-to-read format that increases reader understanding and confidence while minimizing reader fears. Within the pages of *Rescued by Personal Computers* the reader will explore how to:

- Set up the personal computer

- Take the personal computer for a test drive

- Locate and run programs

- Use a word processor, spreadsheet, and database

- Create professional presentations

- Understand the Internet and Web

- Send and receive e-mail and faxes

- Experience multimedia on CD-ROMs and the Web

Rescued by Personal Computers is the ideal book for users who have just purchased a new PC or notebook computer, as well as for the employee whose new job requires computer literacy.

$24.95 USA
Available: January '98

ISBN: 1-884133-54-1
320 pages • 8 1/2 x 10 3/8"

By Kris Jamsa, Ph.D.

> *There's no better way to learn software on your own than to have Kris Jamsa at your side.*
> —Al Harrison, Member of Advisory Board, *PC World*

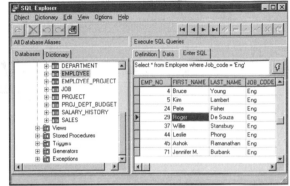

To redeem this coupon, mail the original coupon (no photocopies, please) along with payment and shipping information to:

Borland International, Inc.
Order Processing
P.O. Box 660005
Scotts Valley, CA 95067-0005

Name _____

Address _____

City _____

State/Province _____ Zip/Postal Code _____

Phone (_____) _____ Fax (_____) _____

Select one:

❑ C++Builder Standard for Windows 95 & Windows NT	CD ROM	$99.95
❑ C++Builder Professional for Windows 95 & Windows NT (Reg. $799)	CD ROM	$299.95
❑ C++Builder Client/Server Suite Windows 95 & Windows NT (Reg. $1,999)	CD ROM	$1,699.00

Method of payment:

❑ Check enclosed (Make checks payable to Borland International, Inc.)

❑ VISA   ❑ MasterCard   ❑ American Express

Card number: __ __ __ __ - __ __ __ __ - __ __ __ __ - __ __ __ __

Expiration date: __ __ / __ __

Subtotal $ _____

State sales tax* $ _____

Freight ($10.00 per item) $ _____

Total order $ _____

Offer Code 1479

Offer expires October 31, 1998.

This offer good in the U.S. and Canada only. International customers, please contact your local Borland office for the offer in your country. Corporate Headquarters: 100 Borland Way, Scotts Valley, California 95066-3249, (408) 431-1000. Internet: http://www.borland.com/ Offices in Australia (61-2-9248-0900), Canada (905-477-4344), France (33-1-55-23-55-00), Germany (49-6103-9790), Hong Kong (852-2572-3238), Japan (81-3-5350-9380), Latin American Headquarters in U.S.A. (408-431-1126), Mexico (525-541-1413), The Netherlands (+31 [0] 20-503-5100), Taiwan (886-2-718-6627), and United Kingdom (1-[0800] 973139)

www.borland.com